JOURNEYS WITH THE ICE BEAR

NorthWord Press, Inc.
P.O. Box 1360
Minocqua, WI 54548

Photographer's note: The photographs in this book along with the writing are the
truth of my efforts and profession. The animals are wild and the landscapes are not
computer enhanced. The photographs are real and true.

Designed by Lisa Moore

For a free catalog describing our audio products, nature books and calendars, call
1-800-356-4465, or write Consumer Inquiries, NorthWord Press, Inc., P.O. Box
1360, Minocqua, Wisconsin 54548

For more information on Kennan Ward stock photography and products,
call 1-800-729-5302

Library of Congress Cataloging-in-Publication Data
Ward, Kennan.
      Journeys with the ice bear / by Kennan Ward.
         p.   cm.
      ISBN 1-55971-577-4 (hardcover)
      1. Polar bear.   2. Polar bear—Pictorial works.
    3. Ward, Kennan—Journeys—Arctic regions.
    4. Arctic regions—Description and travel.
    I. Title.
    QL737.C27W375 1996
    599.74'446—dc20   96-13251

Printed in the U.S.A.

JOURNEYS WITH THE
ICE BEAR

TEXT AND PHOTOGRAPHS BY
KENNAN WARD

NorthWord
PRESS, INC.
Minocqua, Wisconsin

Nature is the inspiration and the creativity,
I watch in humble awe of this grandeur and feel
the passion for life grow. . . .

Acknowledgments

Thank you to Karen Ward, wife, adventurer and wildlife enthusiast.

Thank you to the special people who read the first drafts and made a contribution toward the final work: Tom Bentley, Shane Moore, Gerald Garner, Walt Kiernan, Mele Wheaton, and Louis Schnaper.

A special thank you for encouragement and field support: Wolfgang Bayer, Robert & Carolyn Buchanan, Walter & Candace Ward, Thomas & Elizabeth Ward, Betty Ward, The Frear Family, Janis Schnaper, Dave Frenzel, Paul Nader, Mike Maples, Denver Holt, and Chris Watson.

For sponsorship: Royal Robbins, North Face, Willis & Geiger, Lowepro, Marty Stouffer, and Patagonia.

To the kind people in the Arctic villages who helped me in many ways: Usakovski Village, Russia; Arctic Village, Northwest Territories; Churchill, Manitoba, Canada; The friends in Barrow and the North Slope Borough of Alaska, and many other small stops along the ice where friends shared information and a pot of coffee. Thank you!

For enduring friendships: Pasha, Nusha.

I am grateful to the staff at Kennan Ward Photography and to all our loyal customers who have supported our line of products over the last fourteen years. Your purchases have helped us be a voice for wilderness preservation and education and have enabled us to work in the wilderness for an extended time. Thank you!

Contents

Foreword

IF YOU'RE LIKE ME, you've picked up this book because you have an interest in wild animals—especially large predators; if you're like me, you also lead a modern life where in the course of your typical day you do not have to concern yourself with the possibility of bumping into any large predators. Like most people, you probably consider this corollary of modern life a definite plus, but I advocate that it is also a certain minus, because it has left us with a gap: we now miss the sense of awe and the sense of our place within the natural world that, in the ancestral past of our species, were acquired by direct encounters with the few animals on this planet that are one position above us on the food chain.

I think there is a good chance this missing experience is part of the reason you have picked up this book; it is as though we have a totemic link to these animals that spurs our curiosity to know more about their lives.

In our modern world, devoid of direct encounters with these animals, we must now learn about their lives from the naturalists who study them. I divide these naturalists into two groups: the scientists and the artists. Of these two, it is the artists—the writers, cinematographers and photographers—whom we most rely on to fill the gap in our knowledge in a visceral way; their work is what allows us to imagine the look, smell and sound of the big predators. They help us imagine what it was like in the world when in the course of their daily lives people did have to be concerned about bumping into a big predator.

The work of these photographers and writers who record what is left of the wild world is an essential complement to the knowledge the scientists bring back from the field. Knowing that the individual polar bears in the Beaufort Sea can have a habitat range of over 100,000 square miles is fascinating, but looking into the eyes of that polar bear in one of Kennan's photographs is knowledge of a different sort; it is a peek into the soul of the Ice Bear.

Once in the Canadian high Arctic, I was on an expedition photographing beluga whales in the Cunningham Inlet, near Resolute. One morning at about 3:00 a.m., one of our party woke the rest of us in a panic to report a bear and her cub approaching camp. We crawled out of our tents, including our Inuit guide, who had grabbed his rifle. We scrambled to a hilltop next to camp. There, we watched in the Arctic half-light as the bear and cub detoured around camp and continued to the pack ice that choked the entrance to the inlet; there the bears walked onto the ice and into the water.

That was the only encounter I've had with an unhabituated Ice Bear, but I will never forget the mix of awe, fear and respect watching the bear and cub pass near our camp; it was that experience that whetted my interest in Kennan's book. Even if you've never seen a polar bear in the wild, however, Kennan's photographs and stories will give you that sense of awe that comes from being in the close company of the Ice Bear.

Most importantly, Kennan's book will increase your respect for the Ice Bear, and with that respect comes an appreciation that I hope will convert you—if you are not already—into an active advocate for preserving the Arctic habitats of these great animals.

This is the ultimate importance of *Journeys With The Ice Bear*—to remind us that the large predators are the most significant ingredient in wilderness that keeps it wild. Remember, Henry David Thoreau did not write "In wilderness is the preservation of man," he wrote "In *wildness* is the preservation of man."

That distinction is what this book celebrates.

Rick Ridgeway
Explorer, Cinematographer

Preface

To FIND OR WITNESS something special or different about the land, an animal, or about nature has always been the driving force of my work as a professional photographer. In my book *Grizzlies in the Wild*, I wrote about my excitement of being able to photograph clam digging bears, a behavioral aspect that hasn't often been observed, let alone recorded. However, it is possible that most people prefer the more familiar or typical shots of wild bears, like those of a fish leaping into the mouth of a bear at Katmai National Park, or images of the many clustered bears from the world-famous McNeil River State Sanctuary.

I, too, have worked in those popular areas and found photographic success. What seemed to be missing for me there, though, was the more compelling story of the timeless, "undomesticated" encounters with wildlife. Encounters in a place where there are no cabins, viewing platforms, armed rangers, warming huts, or other conveniences. In other words, places where one has to work for the reward of one-on-one experiences with nature and the animals in the wild.

Wild animals behave differently from people-adapted (habituated) animals. For one, they tend to be much more shy and more focused on searching, hunting and feeding, so photographing these animals can be tricky. In general, when photographing in the wild, I find myself in a stalking mode, similar to a predatory or hunting role. This extreme is necessary so my presence is undetected, both when shooting and even when leaving the animal, so as not to disturb it. You never quite know for certain, but if you can observe that you haven't altered the animal's behavior by your presence, you've likely been a successful hunter.

When I began thinking about what would be a unique theme for this book, I remembered my first experience with polar bears in Kaktovik, Alaska. Flying over the native village on the far northern slope of the Brooks range, on our way to a caribou calving ground in late May, Karen, my wife, and I saw a polar bear patrolling the ice five hundred feet below. With a glance, we knew we shared a mixture of emotions, including awe and anxiety.

We were headed ten to fifteen miles up-river on the coastal plain. After landing, we unloaded the 185 Cessna bush plane that had deposited us onto a gravel stream bed, hauling out camp gear for a two-week stay.

Snow was packed along the banks of a nearby stream. Karen and I looked at each other, not needing to say anything: below our feet was a large bear track.

It wasn't until we saw a grizzly bear near camp a few days later that I realized the track had been made by a grizzly. Somehow, the combination of bear sightings added even more anxiety—two large carnivores were within walking distance of our camp.

Our first impressions of the high Arctic were greatly magnified by that brief polar bear sighting from the air. And although we didn't observe another polar bear over the course of our stay, I believe it was the beginning of a new path set out for me.

Slowly, steadily, I began thinking about the circum-Arctic habitat where the polar bear lives. Over the years, circumstances fell into place numerous times to put Karen and me back into ice bear country—where you can't rehearse your photographic opportunities. This book is an accounting of those experiences.

Kennan Ward

The Hunt

THE DAYS HAVE BEEN TOO WARM for any extended activity, so the polar bear rests throughout the midday near an open crack in the ice, alongside a mound of pack ice. All the seals are down in the water seeking fish to eat at this time of day, anyway. Yet, as he rests he occasionally wakes to a smell in the air as the wind shifts. The scent is not strong, but seems familiar—maybe it is from a seal. He drifts back into rest.

As evening approaches, the sun appears to have only rotated around in the sky, not wanting to set. Darkness has left with the last season and the sun will not set until long after the ice has retreated north.

In the distance he can smell the seals that have hauled out now that the sun is in the northern sky just above the horizon. The air is cold again and he feels comfortable enough to move about, as the seal's scent in the air makes him hungry. He smells mostly "tiggak" seals (males over two years old have a musty odor during the mating season), so he must travel to find the young seal pups which taste better.

As he begins his search he finds his frequent "shadow" sleeping in a snow crevice. The Arctic fox and the polar bear have traveled together often. The fox does not always follow the bear to eat what is remaining from the hunt, however. The little fellow is a pretty good hunter itself and is often already eating when the bear comes across a lair. In areas where the fox has marked the seal lairs with scat and urine, the hunting is easier for the bear, but still he is not always successful.

Miles away, on its wanderings, the bear now smells the scat of his fox neighbor on top of a snow bank, and he begins to investigate. Stopping, the bear waits and listens for sounds. Below in crumbled ice lies a birth lair; he listens to the bawl of a pup requesting milk. He hears the exhale of a returning mother at the open water entrance.

He begins to visually map out the seal lair to pinpoint exactly where the seal pup lies in comparison to the open water entrance. That entrance is where the seal will try to escape once it detects his movements. He must trap the seal before it reaches the water hole. He must act quickly. He listens further and his stomach growls in hope of a meal.

The bear's muscles quickly coil and his front paws rear up into the air and pound the ice directly above the lair. A couple of direct pounces break through the ice, pinning the rear flippers of the retreating pup. One extra pounce and he begins to tear at the fur to open its body. The seal's blubber is his favorite food and this pup has turned its mother's rich milk into a thick layer of blubber. Without hesitation he finishes the blubber and, because he is very hungry, eats some meat. Some of the meat, bones and head lie deeper in the snow, compacted from his hunting blows. And splashing water from the lair's exit hole quickly soaks and freezes over the rest of the remains.

The Arctic fox sits nearby waiting for the bear to finish, and ivory gulls and ravens have begun to gather. As the bear finishes, he leaves behind a good portion of the seal, as he knows there are more to be found in this abundant hunting season.

He wanders while the sun wheels around the sky. He keeps traveling and stops only when he smells the fox scent again or, even better, a young seal itself. However, most of the lairs he digs into are empty. Many he passes by are stinky tiggak lairs. He also misses three chances at pups occupying birth lairs.

The sun begins to rise higher as it stretches toward the east and the bear decides to swim across an open water lead to seek juvenile seals that rest on the other side. He plans to catch them as they leave the ice and dive into the ocean on their way to hunt for fish.

The water seems warm to him now compared to the cold wind-blown ice and he feels more at ease. There is no hurry to leave the water, but once again his nose directs him to a far ice edge. This time the distraction is not food; rather, a female in estrus. He follows the scent along the ice.

This is the first bear he has seen in a while. They are both hesitant. He knows she may be dangerous. Then, a grouping of seals lying on the ice distracts her and she slowly stalks one. She lies flat to the ground inching forward; she moves closer, then sits up, snorting from her nose as if she has approached a tiggak seal. The alerted seals quickly retreat beneath the ice.

The two bears hunt and wander the ice pack in close proximity for days, but not too close, just enough so they know of the other's presence. On the third day her urine scent is strong and he finds himself drawn to her. At first a fight ensues, as she tests his strength. He knows it will be his endurance that will prove to be an invitation to mate. After an hour of sparring they couple for over thirty minutes. This is repeated three more times in the two days that follow.

On the third day she has vanished. He follows her tracks and scent, but after a few more days, her scent has weakened, and seals occupy his thoughts again. He hunts now as if his life depends upon this season of seal pups.

BEAR TOWN
Churchill, Cananda

IT WAS MANY YEARS after spotting our first polar bear from a bush plane over the Alaskan high Arctic that my wife Karen and I would again have the opportunity to see polar bears. This second polar bear sighting was in Churchill, Manitoba, Canada, where each autumn the bears congregate waiting for the ice on Hudson Bay to freeze. To see this many bears is a photographer's banquet, an all-you-can-photograph smorgasbord. With this encounter, I came away not just with photos, but much, much more.

Hudson Bay is a unique habitat: a very large body of water that penetrates deep into an otherwise land-locked interior. Its very nature expands the usual definition of a bay—possibly a better word to describe this incredibly isolated water is a *sea*. It is an ecosystem separated from the Arctic and North Atlantic oceans by large islands; Baffin and Greenland lie at its eastern and northern boarders respectively, and narrow entrances of the Hudson Straight to the east and Fury and Hecla Straight to the northwest provide tidal influxes.

In the southwest quadrant of Hudson Bay lies the town of Churchill. It is here that the polar bears come to wait while the ice first freezes. After a long summer grounded on land, the bears wait on the farthest point out into the bay. For when the bay finally does freeze, sometime in the month of November, the bears will again have access to seals, often their first meal of meat since the sea ice melted in late summer.

These polar bears in Churchill and southern Hudson Bay are the most southern populations of polar bear. In contrast, the polar bears of the high Arctic ice cap live on the year-round ice of the polar basin. When the summer break-up begins in April, those polar bears stray north, following the ice as it retreats northward to the permanent ice edge.

According to Gerald Garner, a polar bear biologist with the U.S. Biological Survey, U.S. Fish & Wildlife Service, who has devoted his life to the study of polar bears, "Polar bears are international. They may be on the ice in Wrangel Island [Russia] in the spring and by summer or fall be in North Alaskan waters. They move with the ice."

The movements of polar bears associated with the polar basin are extensive. Put into context with the movements of the ice pack, this becomes understandable. In the North Pacific rim, Bering Sea and Chuckchi Sea, in summer the ice "recedes approximately 870 miles from the maximum ice cover," according to Garner.

The polar bears studied by Garner traveled this ice edge back and forth several times, logging many kilometers of movement. The most movement is in April and May during prime seal-hunting season. Garner has written, "The polar bears in the Beaufort and Chuckchi and Bearing Seas have a pelagic pattern, remaining on the offshore sea-ice throughout the year." These bears do not use or have a need for land. The large ice mass and ice edge is a great habitat for them. The entire polar ice cap rotates, Garner discovered by tracking radio-collared bears. They must move continuously in order to stay in one place, and the ice edge is their prime habitat.

Their large annual ranges easily afford that movement. Garner found that polar bears of the archipelagic habitats of Canada have ranges from 965 to 8,800 square miles. Those living in the pelagic habitats of the Beaufort Sea (Alaska) range 3,800 to 104,000 square miles. And ranges in the pelagic sea-ice habitats of the Bering (Alaska) and Chuckchi (Russia) seas are the most extensive, at 57,000 to 135,100 square miles. As a comparison, in "Grizzlies of Mount McKinley," Adolph Murie found grizzly bear ranges in Alaska to be between 1 and 13 miles in length.

Ice conditions are not predictable, nor comparable year to year, or location to location. Changing ice conditions mean changes in polar bear activity. The summer break-up of the ice in Hudson Bay during July through October (generally considered open-water months) causes the southern polar bears to become land-locked. Patterns of movement change because of this isolation and due to the seasonal absence of ice.

The retreat of sea ice in Hudson Bay restricts the broad movement of bears. These bears respond to being cut off from their favorite food, seals, by changing their activities on land. All bears—brown/grizzly, black and polar—respond similarly to limited access to foods. Minimizing energy while on land conserves body fat added during the seals' pupping season in April and May, when hunting forays of the polar bears are most successful. Staying close to the shore, polar bears cut off from the sea ice find themselves concentrated with other bears not normally encountered on the expansive sea ice. Patrolling the coastline for occasional whale carcasses or other carrion, polar bears in the Churchill area have also been observed consuming sea kelp. These food corridors and coastal land-use cause polar bears to encounter each other more often. This also promotes social behaviors in the bears. It has long been said that the polar bears of Churchill are different, and that seems to be confirmed by research and environmental information.

One additional factor to consider is the contact with humans these polar bears have while being land-based from July through October (and possibly into November). Bear habituation has helped spur an increasing eco-tourism interaction. This active economy is important to an isolated village with little opportunity for income. In the fall, visitors stream in to ride the "tundra buggies" and view the bears. In fact, local travel literature promotes Churchill as the polar bear capital of the world.

Bears standing on hind legs, boxing playfully is a great tourist attraction. Social contact in play or sparring fights is only one of the highlights; what was truly amazing for me was the seldom seen polar bear "pile." On a night excursion I saw a group of bears piled together in protection from a snow blizzard. I called them "sugar bears," as they were coated with light fluffy snow. Piled up four deep and more, the bear piles consisted of mothers with cubs and groupings of juveniles and adults together. Clearly these reportedly solitary animals can be very social.

Photographers and cinematographers often use Cape Churchill, located about 60 miles northeast of the town of Churchill, for their work. Each November, a small group of people live for ten days to three weeks in "tundra buggies" engineered to endure the cold. These are designed, assembled and maintained by local resident Len Smith, a creative and gifted mechanic. The vehicle's large tires—five feet in height and four feet wide—were adapted from agricultural use. The original chassis was derived from a garbage truck, topped with the body built by Smith.

These all-track vehicles pull sleeper cars, a kitchen/dining car and the essential tool shed/generator trailer. Besides photographers and working media personnel, some brave eco-tourists and avid amateur photographers make their way out with this group. Doctors, lawyers, professionals and their companions round out the group. This guided "soft adventure" was the first group activity Karen and I had experienced. For a few guests the persistent cold, extended bear watching, and the limited quarters were beyond their enjoyment threshold. In contrast, the enthusiastic at heart marveled at the absurdity of a hot shower (or lack thereof) in the wilderness Arctic; and for photographers, the time spent was very rewarding with prolific results.

Many photographs of spirited bears have been composed here, as the seemingly carefree bears await the freezing of the bay. As the hours of sunlight shorten daily, the tundra vehicles patrol the coast in search of polar bears. In November the sun rises after 8:00 a.m. and sets shortly after 4:00 p.m., and actually rises only to a low angle in the sky. This pattern generates good photographic light conditions practically all day. With the abundance of reflected light off the snow and ice, even during storm conditions (when experimented with) can produce good photographic results.

To focus in on a subject like polar bears is a great opportunity for learning and developing a deeper understanding of nature. A group of people brought together and stuck together out of devoted interest in polar bears is a study in itself. Len and Bev Smith have done this longer than anyone, and yet there remains amazement in their eyes and fun in their smiles as they watch the great ice bears. No cameras in their hands, just pure delight at witnessing the world's greatest polar bear show on earth, "Bear Town," Manitoba, Canada.

On the Edge at Lancaster Sound

ON ONE OF OUR FLIGHTS to polar bear country I began to ponder how to describe what it is like for Karen and me to travel to a place like the high Arctic.

Perhaps *journey* is the best word. For us, journeying is traveling for a long distance over a long period of time, a month or more, to an often unheard-of place. Spice that recipe with hazards like "mega-fauna," sub-zero temperatures, occasional 40 to 60 mile-per-hour winds, floating-ice camping—it makes for a whole list of exciting journal entries.

One of the best illustrations of a typical journey was our experience with exposure to the conditions under which polar bears live. What these animals do as routine is death-defying to humans. The wanderings of the ice bear are difficult to trace, and to fathom. Take a set of tracks along the ice floe: up out of the water onto soft ice, then down through a seal hole and out again along a crack in the ice. Miles and miles of distance traveled through a harsh environment, around the moving ice edge.

In this unique environment, extremes seem ordinary and adaptations develop. Summer is the time of the midnight sun. North of the Arctic Circle, between May and August, the sun stays above the horizon. During the winter, begining in December, as the sun nears the horizon, the long hours of night are followed by a few hours of twilight. North of the Arctic Circle, in winter, the sun does not rise above the horizon for many weeks. It is during this time that the Northern Lights, *Aurora Borealis*, may be seen on a clear cold night. I often wonder if animals have an awareness of this phenomenon or are affected by its ionic reactions.

Our knowledge of this large wandering carnivore of the Arctic—called polar bear— had been limited to mostly cursory observations so far. Several factors had challenged our contact with and understanding of these elusive animals. In the past, difficult obstacles to long-term observation have included severe and variable weather, changing sea ice, and the bears' ease of swimming through the waters along the ice pack, making it impossible for humans to follow tracks.

In fact, during our first week camping out on the ice edge of Lancaster Sound, Karen and I didn't see bears; tracks were the only mark of their presence. At one point we noticed a large track in powder snow, the rear imprint showing only a ball shape where the heel touched the ground. Missing was an impression outlining the entire foot. Our Inuit guide said, "This bear is hungry." The bear's step was not heavy; otherwise, the powder snow would be showing a fully shaped track.

Along a trail of paw prints left on the ice, I came upon a large scat (excrement). Its size proclaimed that it could be from no other than a polar bear. Seal fur made up a portion of this dropping, evidence of polar bears' favorite food. Further along, the bear tracks crossed a set of wolf tracks. Or did the wolf cross the polar bear tracks? The clue was a track deep in the snow near the bear tracks, suggesting a stop in the wolf's movement. The prolonged warmth from the wolf's paw had melted more ice and snow, creating a deeper and slightly wider track. We could see a nose print within the impressions of the polar bear's toes. The wolf had examined the smell left behind in the polar bear track. The tracks of the wolf and the polar bear paralleled for miles after this intersection until the bear tracks led to a seal lair.

From afar we saw some color in the monochromatic, white landscape. On top of a section of pack ice a pattern of blood surrounded a hole piled with snow trailings. The hole and diggings went down to the hard blue ice layer, where the remains of a seal pup had been re-frozen in the sea ice. The wolf tracks read as if it had come by long after the kill and investigated as we had.

Undoubtedly before the arrival of the wolf (or us), following closely behind the polar bear was its shadow—the Arctic fox. Arctic foxes survive partly off the leftovers of their larger and more successful predatory neighbors. Bears prefer the inner skin and blubber of the seal kills, while the Arctic fox finishes the meat. Ravens and Ivory gulls pick the bones clean.

Now our tracks became an addition to the story. The bear tracks departed from the seal kill and were followed by or paralleled the tracks of the Arctic fox, wolf and us. Looking out over the ice, the tracks extended for miles, toward the horizon, to the floe edge. When we reached the floe edge, the polar bear tracks disappeared at the edge of the ice, pointing toward the water. The wolf tracks stopped at the edge, forming a deep-tracks "portrait" of a loitering creature, much as my tracks appeared as I lingered, imagining the unreachable realm of the polar bear in its marine environment.

The wolf tracks, Arctic fox tracks and human tracks all followed the floe edge, now in the height of the rose-colored midnight sunlight. The floe edge is the activity center of the whole ice sheet. The next morning I discovered that the bear had come back to our side of the ice. After only two or three steps it had shaken itself dry upon the ice edge. Water droplets perforated the snow in a heavy shower of scattered drips, spreading out to a light sprinkle farther away. Frozen puddles formed beneath the tracks, as the water must have poured off the bear's belly fur. The tracks then followed the ice edge toward our camp. But these weren't the shallow rear paw tracks we had observed earlier. The whole rear pad was left as an imprint upon the hard snow and ice, even after the water's influence. This was another bear, well-fed and heading toward our camp.

These tracks—clues and mysteries left behind—helped us to understand the behavior and the movements of the polar bear. This kind of shadowing always provoked deep anticipation and excitement in our journeys, feelings that became even more intense as tracks and sightings began to accumulate, as if I were reading the signs of several mysteries all at one time, and that if I put together all of the clues I would understand the heart of this powerful animal.

The endless daylight of the Arctic intensified by the ice-bound landscape adds drama, while the cold and ever-changing weather and ice conditions challenge your imagination. We found our imaginations were never short of challenges, and that the distinctions between the tracker and the tracked are often blurred in the whirling Arctic winds.

><+>—o—<+><

Also during our stay along Canada's Lancaster Sound, we often looked for polar bears from a ridge overlooking the ice floe-edge of Dundas Harbor on Devon Island. Although it was only mid-May, the open ocean lead (crack) in the ice was already as close to shore as it would normally be in June. The large lead changed constantly as the ocean tides, currents and winds moved the unfrozen body of water below. The upper frozen layer of ice moved in large chunks resembling islands, connecting together then breaking apart.

The newly exposed water released a fog into the air, then it turned to slush, then froze into mirror-like ice. Walruses broke through the slush and thin new ice to breathe and Eider ducks walked upon it in search of open water. Beluga whales and ringed seals also passed through an open water corridor that provided breathing holes in the mass of endless sea ice.

The warm hues of the midnight sun changed to depressing grays as we finished a ten-hour watch for polar bears. An approaching dark cloud swallowed the 3,000 foot mountain to the east where we had spotted musk oxen. Wind scraped across the frozen harbor beneath our overlook. Denied the little warmth the sun had provided, the added wind chill began to burn our lips, cheeks and noses with every inhalation. No bears could be spotted.

Packing up our equipment became an event. Gear management in cold blowing snow is one of the scenes most people don't envision when they imagine the life of a nature photographer. These situations usually provoke our attempt at comic relief with lines like: "This looks like a page in the book, *So You Want to be a Nature Photographer.*"

Blowing snow filled the lenses as we quickly disconnected cameras and packed them away. Snow entered everywhere as zippers froze open or pull-tabs broke loose in the icy struggle. Numb fingers, robbed of warmth through contact with metal fasteners, strained to loosen tripod legs and knobs. The wind blew my hat off and I ran after it sliding into an icy ditch, catching the hat as I fell. Finally, with all the gear collected and packed, we began the hike down the hill.

From our overlook, a trail of rock and ice led to a qamutiik (a wooden sledge pulled by a snowmobile), which would transport us and our gear back to camp across the ice. The qamutiik rested on the shore of a solid ice harbor leading to the head of a beach several miles north, where our camp awaited its first storm. Carrying four packs of camera gear weighing fifty to sixty pounds each down to the qamutiik was a routine chore.

At one point, a 40 mph gust kicked at the large duffel pack on my back, rotating and twisting my body while it corkscrewed one of my feet into the icy snow. Then the gust whipped away as quickly as it started, motivating me to speedily finish two more pack trips to the qamutiik. These routine procedures are not particularly memorable. What we do remember is the incredible scene of the ice floe closing as the winds brought the sea ice together, snapping shut breathing holes for the wildlife.

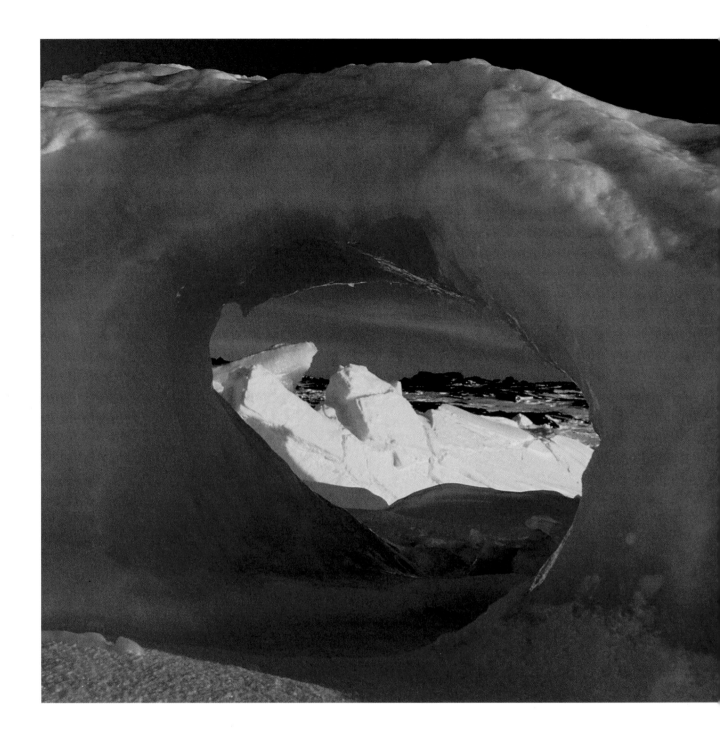

The ride to camp across the ice was another experience in itself. Blue icebergs frozen into place were surrounded by green pools of water where the high-tide waters surfaced through the surrounding ice. We traversed large cracks—of seemingly endless depth—a foot or more wide. The ice shifted, zinging and moaning, as the sled rails skimmed across the cracks. Our senses suggested the ice was alive. We skimmed past a large iceberg that hosted a three-foot circular hole in its huge mass, framing a view of our look-out hill.

Back at camp we prepared for the wild storm by reinforcing tent stays and gathering gear. As the winds blew and gusted we finished our chores and prepared a meal. Just as we began to eat, we spotted a polar bear about a mile out of camp, out on the ice in the middle of the harbor!

The mile that separated us seemed so much farther due to the blur of blowing snow. Behind a bluish iceberg was "moving yellow snow," a name given to polar bears in winter. At one point the "yellow snow" separated into a large piece and a small piece. The blowing snow subsided for a minute and a mother polar bear and her three-month-old cub became clear.

The mother bear and cub had known about us long before we had seen them. She had been waiting and watching from the side of the blue iceberg she'd selected as a shelter when the storm first began. The leeward side of the iceberg blocked her young cub from the wind. From this protected angle she could better sense what we were, analyzing the smells and sounds in the air, and what she saw in our movements.

Karen and I responded to the bears by preparing cameras, lenses and tripods. I grabbed an 800mm lens and set up on a tripod close to the ground. The wind shook my lens so much that even if the bears were to come close, a picture would be impossible.

After a quick look around, we headed to an abandoned Royal Canadian Mounted Police (RCMP) post work shack that we had seen earlier and took cover. We set up inside and pointed the lens through the broken window. We waited and watched as the winds temporarily subsided allowing the mother and cub to approach. Slowly, they moved with the wind toward our camp, watching, the sow occasionally lifting her nose to test the air for scent.

Watching the small cub walking across the ice in the blowing snow, I thought of how much of a struggle it had been for me just an hour earlier as I retreated from that same location, in the wake of this storm. How fragile the cub appeared against the hazards of ice, snow, open cracks, the cold, the wind and the unknown: us.

The mother bear approached closer, cautiously surveying the route ahead. At one point she stood with her nose high in the air trying to catch another scent in the wind, or perhaps to better detect motion from a higher perspective. The standing bear with cub to her side was more a picture for the mind than a literal polar-bear-in-a-snowstorm photograph. We attempted a photograph anyway—this was our first bear sighting through a camera lens in more than a week of effort.

The young cub's adventurous journeys onto the ice demonstrated the unmatched skill, strength and survival of the largest land carnivore. I was left in awe, and I remain wondering and curious about this brave animal, hoping to learn more.

The polar bears traveled farther downwind, finally reacting in distrust to the scent from our camp. Immediately the female stood, then broke into a run, slowing only to allow her cub to catch up. Within minutes she was more than a mile away heading toward our hilltop overlook; moments later she was out of sight. Whatever she thought we were, she had made a move to protect her cub and herself from further contact.

This bear was completely elusive, showing no curiosity about exploring a human scent. The stories of polar bears stalking humans, ready to terrorize, or as the subjects of haunting imaginative stories didn't seem to apply to bears on the ice, who spend their critical time hunting, feeding, and surviving the extremes. When the occasional sighting was made of a wild bear, they would instantly run away when approached within a couple miles. I worried that I would not be able to illustrate these truly wild animals with photographs, these polar bears at the floe edge, in their private wilderness.

The possibility of working with polar bears along their hunting routes on the ice became even more difficult as later in the evening, after the midnight sighting, the storm built in intensity. A canvas tent collapsed in the winds, crashing down upon fellow camper and renowned cinematographer Wolfgang Bayer. He was struck in the head by its 2 x 4 center post. He pushed the pain aside as he spiritedly joined in the chore of refitting the tent and steadying it against the growing winds.

Our four-tent camp was now seriously threatened by the building winds, blowing at 40 to 50 miles per hour with occasional stronger gusts. For the entire morning and most of

the day we continued to reinforce the tents against this tireless storm. I wondered how the female and her cub were coping with the wind, or if they even noticed.

The next morning, Wolfgang, his assistant Ian and our local Inuit guides Johnny and Abraham retreated from their collapsed tents to an abandoned RCMP building. During the late morning, they reestablished camp inside the building's largest room. Debris and snow were shoveled aside and the broken windows covered by tarps. The tent camp had been blown apart and away. Gear, debris and supplies whirled madly in the wind's direction.

I was hit at knee level by frozen packaged meals blown loose from their box; there were empty gas cans orbiting on the wind, along with everything else that wasn't tied down. I made my way back to the only tent standing, with Karen sheltered inside.

Karen and I took turns reinforcing our tent against the storm, with one foot against the interior side bracing against the cross poles. In this manner one of us was able to sleep for a few minutes at a time. After several hours, I went outside to check on the RCMP house. The four men were settled in for a much needed nap. I could see the heavy wind had taken its toll on the camp. The nearby ice was littered with stray equipment and goods.

Midday brought round two of the storm's high winds. Eighteen hours into this storm, we planned a retreat into the work shack we'd utilized when the polar bear and cub came by at the beginning of the storm. We loaded our tent and gear onto a qamutiik and hauled it to the shack. Karen shoveled snow and dirt out of the interior and uncovered a solid wood floor in remarkably good condition.

We set up inside the small building, just wide enough to accommodate the width of our dome tent with a narrow corridor to walk by. The length of our new shelter was twice the length of our tent. The door and window had been lost long ago. It was nearly midnight when the storm hit again, and its strength shook the building—we shook with the vibrations of the floorboards. We were glad to have the strongly built shack to protect ourselves and our tent, but the noise and ferocity of the storm kept us awake and on edge until exhaustion lured us to sleep.

I listened to the wind blow at our shelter, while thoughts arose about the polar bear cub. I wondered if the young cub was safe. I could hear pieces of ice and snow being uplifted and moved along with the wind. Icebergs shifted in our frozen bay as the ice began to pack against itself. I wondered how the mother would protect her young from these hazards and this storm.

The next day, I wondered all these things again, as our plane lifted off, taking us home.

High Arctic Habitat

I HAVE OFTEN WONDERED ABOUT the appearance of the circum-Arctic landscape. It seems desolate and bleak, but this harsh and frozen landscape holds the magic of life uniquely adapted to this region.

Trees cannot survive in the region lying above the Arctic Circle—not even the tiny spindly black spruce that bravely survives south of the Arctic Circle in regions of discontinuous permafrost. Continuous permafrost is found north of the Arctic Circle,

a layer of ground that has remained frozen for two or more years; the frozen layer in some areas runs to a depth of 2,000 feet. This region of permanently frozen subsoil, long winters and short growing seasons limits plant life to wildflowers, sedges, cotton grass and dwarfed shrubs in moist areas. In higher elevations and drier areas, low-growing tundra plants hug the ground. Within the first few inches, a tiny micro-climate provides enough warmth for slow, limited growth.

Stunning wildflower displays provide a highlight of summer color. These tiny plants, along with lichens and moss provide important forage for wildlife. It is the ice that remains the dominant influence in plant or animal evolution, and is central to their survival in this habitat.

⊷⊶⊙⊷⊶

Gliding across the ice during our trip to Lancaster Sound, Karen and I realized that ice strength was a real issue. It becomes vital when your camp is lying on a surface of ice with a thickness varying from one foot to ten feet—ice which is floating on top of water as deep as the Grand Canyon. We were also well aware that our remote camp sat atop an ice sheet that had strong ocean currents moving beneath it.

The ice issue intrigued me. Some "ice facts" I knew instinctively. Some things I learned when I was a Park Ranger. I had also become familiar with mountain and glacial ice as a climber-mountaineer. In all my experience, however, never did I see or learn about ice as much as in the High Arctic. For instance, there are as many names to describe ice as there are different kinds of ice and natural features connected with ice—pack ice, sea ice, fast ice, ice fog, drift ice and ice out, to name but a few.

A "springtime crack" in the ice is narrower than what is described as a "lead." Leads are large bodies of open water in the ice, longer than cracks and sometimes connecting to other leads. Near our Lancaster Sound camp, we were following a crack that was five feet wide. Fresh bear tracks, marked with dripped water frozen in the snow, traversed this ice crack toward the open sea ice, more than five miles away from land.

The crack ran beyond our sight in both directions but was refreezing as it slowly widened. The ice was reforming from the edge of the crack toward the middle, where slushy water appeared. Without warning a head popped out of the crack and shook. To our amazed surprise, here in the backlit midnight sun appeared a polar bear.

These bears fascinate me, they are so cool—chilling, in fact. Who knows from where this one came or what he had been doing? Had he seen us earlier or did he just surface from a long swim under the ice? Bobbing in the slush, with a paw on the new ice forming along the crack, he watched us for a while as we remained moderately still. It was a reasonably large bear, without cub, so we assumed it to be male. He disappeared under the ice, and we quickly set up camera equipment. Ten minutes later he surfaced a hundred yards north, only his head visible as it shook. Close to the open water, breath vaporizing, he looked at us again as he bobbed up out of the water. Then he went down under for five, ten or fifteen minutes more and surfaced again closer to us.

Our imaginations can only approach an understanding of his world, blocked from our view beneath the ice. The demands of the weather force us to remember how fragile we are so far in the north. The bear has long been adapted to its environment and is comfortable in or out of the water—in the same environment, we hang on to our thin edge of survival. The ice bear's strength, ability and skill is beyond that of any mentor I have had. Having a vocation that relies on wilderness skill, survival, risk and adaptability, I can only envy the bear, as I see him interact with the ice world that would kill me from exposure in minutes.

The polar bear continued to follow the crack ice in a northerly direction, occasionally surfacing for air, looking around, and then disappearing for minutes at a time. When he was about a quarter-mile away he popped entirely out of the water, shook the sea water off his fur, and ambled north along the ice edge. After a few photos of the bear retreating, we followed. His tracks led to another shake spot, and farther away a roll and rub on a pile of snow and pack ice.

Following the tracks was our only trail clue, as he had quickly disappeared behind mounds of pack ice. "Moving yellow snow" polar bears are dubbed, but we could

not see him. Was he between pack ice mounds? Back in the open water of the crack? Swimming under the ice? Regardless, we had lost him. Karen, our Inuit guide Johnnie, and me; three pairs of eyes scouting in three directions lost visual contact and later, lost even the tracks of this bear as well.

The wild nature of the polar bear is far different than that described in the stories of habituated town bears, or human-food influenced bears. The wandering nature of these bears became apparent as we observed them patrol the ice from ice crack to seal hole, presumably on the hunt. We found signs of success—blood at a seal lair—only twice in three weeks.

The constant movement of the bears we saw along this floe edge contrasts sharply with the waiting behavior of bears at Churchill, where they rest until the bay freezes. Or that of polar bears in Barrow, Alaska, that lie watching a whale carcass.

Or even of the slow-moving mother bear resting just outside her den in Russia. What they do have in common is the ice—a small cap of polar ice that lies high on top of the North Pole in the circum-Arctic, where the ice edge moves north in the summer and south during the long winters. It is on this edge of polar ice, this edge of life, where we tried to unravel the secrets of the ice bear's High Arctic habitat.

Beyond impressions of cold, when people think about the Arctic their thoughts are often of a mundane vastness of ice, miles and miles of sameness, like an empty desert. We soon discovered, however, that there was much, much more right before our eyes.

Take the "moving rocks," for instance. We had read that on Canada's Devon Island there was an original population dating back before the Ice Ages, making them an "elder" in the terrestrial mammal kingdom. On our first night at Devon Island while spy-glassing from an overlook hill we saw these far off "moving rocks." From a distance, Arctic detail is often indiscernible—all that we could see was a contrast to the snow, a movement, a curious, ground-level flow of fur. When approached, the dark-brown blackish creature looked like a pygmy tundra buffalo, as once thought by early European explorers. Rather, it is a northern relative to the goat: the musk ox.

It would be very easy to miss this animal if you didn't specifically look for it, or know that it inhabited this niche. Musk oxen are shy and normally slow-moving, and resemble large rocks against this snowy and wind-exposed backdrop. They carve out a living in this severe environment, finding their food in a habitat little else survives in. Even when we were close to them, we were uncertain where they were until they moved. When our eyes adjusted to the minimal contrast, detail became discernible, and when the sun dipped toward the midnight twilight, a photograph became possible.

Also hard to spot are beluga whales—white whales against sea ice. We would hear their exhalations, and that triggered a reaction to look for them as we watched the floe edge. It was our ears that gave direction to a visual sighting. Then one after another the whales would surface to breathe, in the midst of where we'd tracked their sounds. Once you discover the beluga, you may see eider ducks in flocks take wing at the disturbance. Ivory gulls fly in to investigate, and all of a sudden what was quiet, vast and seemingly empty becomes alive, wild and interesting.

Another vital piece in the understanding of this unique community is the walrus. Legends place polar bear and walrus as ancient ancestors. Polar bears were referred to as walruses who evolved from life in the sea water. This belief may have come from observations of walrus; when they haul out from the cold waters they are a light

cream or gray color. During feeding forays, which may take them out to sea for 6 to 10 days, they endure long exposure to cold and their blood leaves their body surface and pools deep into their body core to keep vital organs warm. Their outer layer of blubber insulates their inner core from the frigid sea, and it is this process of the blood not circulating through their outer blubber that turns them lighter in color.

When they haul ashore following these long stints at sea they seem to move with exhaustion, with labored breathing, slowly rolling up on the ice floe or beach with a lumbering, head-bobbing pace. As they begin to warm up, their blood rushes to the surface of their hide, blushing it pink, and gradually turning it to a reddish brown as they warm fully. Either the light gray of the young walrus or the light color of the cold, sea-worn adults could cause these creatures to be mistaken for a polar bear when viewed from a distance.

Adult male Pacific walruses may weigh 3,000 pounds. A walrus or polar bear is a remarkable prize and are eagerly sought by native hunters. Both species are totally utilized by the hunters, and both are sought with equal need.

According to a story I've heard, a group of polar bears jumped onto the backs of some walruses that were a part of a herd resting ashore. One polar bear even held onto a walrus with its teeth and paws gripping and slipping along the back of the walrus, as the huge animal, two-thirds larger than the bear, lumbered off in panic. The entire walrus herd rushed for the safety of the sea. Ironically, it was the walrus' own doing that trapped, trampled and crushed a few members of the herd in their stampede. The meal provided to the polar bear was not a direct kill—the walrus hide was too tough and their round profile made it hard for the bears to get a firm grip. But repeated stampeding of the herd and checking for casualties is evidence of the intelligence of these great bears.

A loud thundering sent our eyes again to look for movement, this time from an iceberg set free by the rising tide and a crack in the surrounding ice. The roll of the iceberg sent a ripple reaction through the slush ice and awakened a walrus resting halfway up the edge of the same ice. The walrus was holding onto the ice above the water asleep, with its tusks embedded as an anchor.

Many times people have asked me what walrus use their tusks for—now I have an answer, based on field observation. We also photographed the walrus using its tusks to climb out of the water onto an ice floe.

⊳–⊦–⊷–○–⊶–⊦–⊲

We easily concluded that in the High Arctic, ice is the all-encompassing habitat. It is unique and vital to many species. For the seal, the most important ice is the stable first-year ice known as "fast ice." They live in the world beneath its surface, and only rest on its surface.

Our camp was on the fast ice at the floe edge. A ridge of pack ice elevated us 100' for observation. With binoculars we could see a long distance on the endless ice plain. Sparsely spaced across the one-to-two mile wide strip near the floe edge were ringed seals lying on the ice.

The ringed seal is abundant in the High Arctic. They are often found in open water at the floe edge, mingling with other ringed seals.

As a marine mammal that breathes air, its path beneath the ice must be matched with its ability to hold its breath. As I looked closer at the fast ice I discovered little holes in the surface, called "breathing holes." As a mark of recent activity, some of them were still filled with open water; others were covered up by re-frozen ice.

I envisioned the ringed seal playfully foraging the water-world beneath the ice, traveling from one breathing hole to another, scratching out a new breathing hole as it passed a long stretch beneath the ice. The ice can measure from paper-thin to many feet in thickness. Seals use their flipper nails to scratch at the ice and puncture new breathing holes at thin places. Flaws in the ice also provide openings

that are utilized as sources of oxygen and are maintained by the seals blowing air and nuzzling to prevent the water from refreezing.

Some breathing holes become expanded with use, until a seal can move completely out of the water onto the ice. These are called "haul-outs."

We set up remote cameras in hopes of photographing these creatures. But every haul-out site had produced nothing but photographs of blowing snow. The skittish seals had eluded our photographic efforts.

Wildlife gather on the floe edge where food is concentrated. As all flora and fauna of all habitats are interrelated, so are those that make the High Arctic their home. It is not surprising that all their nutrition can be traced to one source: yellow ice.

Yellow ice develops underneath the pack ice in the sea water where phytoplankton attaches and grows on the submerged ice layer. In the Arctic storms the ice can be broken and eventually heaped up onto the surface pack ice, allowing us a glimpse at the web of life beneath the surface ice. This algae biomass feeds 200 plankton species, that in turn feed small fish and larger fish, meals themselves for seals, eider ducks, sea gulls, and beluga whales. Walruses feed on mollusks, clams, and fish. The phytoplankton is the base of the Arctic food pyramid, and the polar bear, being the top carnivore—along with humans—is represented as the top of the pyramid (or iceberg, if you will).

During our various trips to the Arctic, we have learned much about this austere but beautiful landscape and its unique inhabitants, and how weather affects them.

One particularly interesting thing we observed is how much polar bear activity levels change throughout the seasons. During the late summer and fall some polar bears are land-locked away from their prey, like those at Churchill. These animals patrol beaches in hopes of finding a whale, seal or walrus carcass. During their patrol, some bears will consume beach kelp that has washed up on shore. Most adjust their behavior and activity to the absence of food, just as a grizzly or black bear responds to winter and the absence of fish, berries and vegetation.

Hibernation is one adjustment, though extreme, to cold and the absence of food. A hibernating animal's metabolic rate, heartbeat and respiration can drop up to 50% of normal. Bears in general are not considered true hibernators because of the ease at which they may be awakened in the den. Polar bears are the least likely to be considered hibernators since apart from maternal dens and holing up for storms they are generally active in winter.

As the ice returns in November and December so does easier access to seals, the bears' favorite food. The seals move with the ice edge and build their lairs a mile or so from leads or "polynias" (a consistent open water habitat unique to the Arctic). The bears are aware of these large niches of open water that provide refuge (and even homes) for traveling seals and birds. During winter, polar bears wander freely, but food is not uniformly available. Polar bears experience extremes in food availability one season to another but there are times when food is found only after broad wandering, and after many failed hunts.

I was greatly surprised by the contrast I observed between the relatively lethargic movement in summer and fall, and the aggressive wanderings of spring. I had no idea these animals worked so hard to obtain their food. They may patrol a thousand miles, seal hole to seal lair, with only an occasional success. We were overwhelmed by not only their constant movement, but also by the energy expenditures required to run, swim, climb and dig through the pack ice.

As an example, in the Arctic I was routinely reminded of my muted sense of smell. My own olfactory sense was lacking. In the cold, scents remain contained—even the smells of food were isolated to close proximity to their source. In warm weather, waves of odor radiate away from an object. Polar bears relying on their heightened olfactory senses have uniquely adapted to their environment, especially as their prey is often out of sight beneath the snow, in a lair.

To learn the ways of the animals in the Arctic one must be a great detective. Many clues are written on the land in the form of tracks. And many are discovered by observing the intertwining of their lives.

Although I didn't know it at the time, all these discoveries would prove to be invaluable in my greatest journey with the polar bear yet to come.

Wrangel Island Expedition

IN COLLEGE I HAD READ about the ice shelf along Arctic waters and the abundance of food and animals. The Bering Sea, the North Sea and the Chuckchi Sea were often cited in scientific journals for the proliferation of available nutrients and the variability of wildlife. Mentioned were animals unique to northern waters, such as polar bears, narwhals, walrus and various seals. And migrants such as waterfowl, California gray whales, killer whales and beluga whales.

Although not much was known about the Chuckchi Sea off the northeast coast of Siberia, it was documented as having water currents that circulated around an island named Wrangel. This island was often cited as supporting an abundance of wildlife, but little in-depth knowledge was available because of its remoteness and the difficulties of scientific exchange with the former Soviet Union. And while Wrangel Island was often mentioned throughout my academic reading, it was not referred to as consistently as Alaska or Canada, both of which were within reach of possible visits. For a child of the Cold War, Russia was not approachable and going there was almost unimaginable.

Until, that is, the cold winter day I received a phone call from Shane Moore—a good friend and cinematographer. He and I had shared ideas, field time and experiences in Alaska for several years. At our first meeting he and his wife Lybby were in Denali National Park starting a moose film project. Shane is a rare field partner, and like one of my ranger buddies, could be counted on for many skills in the wilderness.

He called to say he had a contract to work on polar bears. Without hesitation, I stated: "Count me in on anything." After writing *Grizzlies In The Wild*, I was seeking a new project. It was humorous that a new subject—polar bears—had picked me, although most of my friends had expected something with "bear" in its center. I could not have been happier with this new topic.

Shane could handle himself in bear country alone, but why not have a buddy along? I would pay my expenses and hope the old "safety in numbers" rule would apply. Then the real news came that denning behavior on Wrangel Island was being considered as the subject. To this day it sends a chill (maybe because the entire trip was the coldest time I have ever spent) as the most fantastic journey in which I have partaken. This journey sidesteps the category of the most fun journey, and it widely sidesteps that of the most productive journey, yet it stands worlds apart in terms of photographic perspective and personal evolution.

Primarily based on Shane's recommendation, I was granted inclusion into the Wrangel Island project. Fees were paid on permits, charters and equipment so that

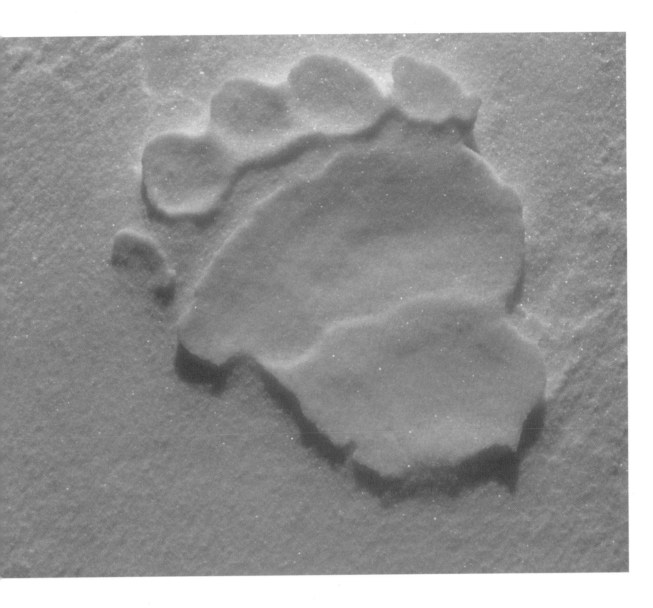

the original cost structure could accommodate two people. I paid for any additional overhead. In turn, I would provide some photographic images for use on Shane's project. A three-way win-win-win: no additional costs for the original project, a safety partner for Shane on a freezing cold expedition, and an experience of a lifetime for me—a journey to a remote, long-sought island with a trusted friend.

What lay ahead was more than I could have imagined. There was a lot more to the puzzle than just California to Anchorage to Nome to Providenya to Cape Schmidt to Ushakovska to Warring Camp! Weather and the lack of basic traveling infrastructure in the newly opened former Soviet Union hadn't been factored in. A three-day snowstorm forced an overnight in Nome, Alaska, an unscheduled delay. The storm was thought to be ending, but radio communication from Providenya, Russia said that it might take several days to clear the runway.

To our delight, by late afternoon on the next day we loaded a Navajo twin-engine airplane and flew west. Once we passed into Russian airspace we lost another day, this time by crossing the international dateline.

⊱──⊰⊱○⊰⊱──⊰

There are numerous pages in the Wrangel section of my journal for March and April 1994 that summarize the efforts of getting to a polar bear den. Here are a few:

Sunday, Monday March 27, 28

Clouds encompassed the entire coastline of Russia and the pilot searched for a break in the clouds. Once he found a small window, we flew in a corkscrew spiral down, and ice became visible in the bay. Landing was smooth.

During our taxi in we passed a military aircraft flanked on both sides with helicopters, fighter jets, large transports and a couple tanks. Everything was snow-covered including the planes and equipment. As we parked, several border guards with rifles surrounded the plane and directed our movements. These were young men, no more than eighteen. In heavy wool coats and large hats, their faces were the only exposed skin. The Russian red star decorated everything. We were in Russia, and this greeting seemed to be what was expected. In the course of all my travels this was an incomparable experience.

Immigration and customs took most of the afternoon, and I think they let us go only because it was almost time for them to go home. Biologist Gerald Garner is our traveling companion. His scientific gear was locked up in a storage area pending further discussions, paperwork and who knows what else. There was a certain formality about the officials. When I was asked to show my currency it caused a scene. Off came my money belt and out from a zipper came hundred-dollar bills to pay for the expedition.

Not only did everyone laugh, giggle and smile but there was even some joking in English. Everyone's human side had come out, and after that smiles and casual interactions followed. The ice was broken for everyone except Gerald, stuck on the side listing every single luggage item onto a tablet. Even his two customs agents had laughed when the jokes about the money belt flew, and in the end everybody was friendly, until a "three-star epaulet" man came in, and then all faces held stolid expressions. He waved me through and escorted me outside the airport, where a couple military vehicles and several army personnel stood guard. I was placed in an old van and felt a bit estranged. Shane followed about forty minutes later and as for Gerald, well, that was a long story!

It is hard to describe Siberia: it has some similarity to other Far North civilizations and other aspects that are unique to Russia. Waiting and vodka are about the only two things you can count on in eastern Russia. Because of the weather, we became stuck in Providenya for a week.

We were all aware that this delay was becoming an opponent in a fight against time. Stranded in this town we were losing valuable time, time when the female polar bears would be emerging from their dens with cubs.

Upon finally flying out I wrote:

This place has provided enough material to write a non-fiction work; this day and the week prior underscore the extremes to which we have been exposed in our search for polar bears—and we aren't even on the ice yet!

Nevertheless, we survived and finally a plane arrived from Cape Schmidt with Misha Stiashov, a biologist at Wrangel Island Reserve. A nine-cylinder rotary engine powered the bi-winged aircraft.

Beginning after a 6:00 a.m. breakfast, it took until 3:40 p.m. (twenty minutes before the airport closed) to clear customs, immigration, border guards (former KGB) and miscellaneous loops, hoops and bribes, until we literally escaped with eighteen-hundred pounds of gear, fuel and people. At that point, if we didn't just leave we would have been detained there at least another day. In a matter of minutes the bi-winged plane lifted off the runway like a smooth elevator, slow and steady. I looked out of the round window to the frozen volcanic landscape below. There was only one seat in the back; I sat between boxes and camera cases.

The plane's cargo consisted of unsecured gear, fuel drums and aircraft hardware. The crew of the aircraft included two senior pilots (Roman and Mikhail) who seemed to debate most of the flight procedures, and a mechanic (Vladimir). Once we were

airborne Vladimir passed around an eight-liter jug formerly utilized for engine oil, now filled with Peeva, a Russian beer. I was told that it was custom and tradition to drink and an insult to refuse.

I look below and see mountain ranges of volcanic domes and coastal plains leading to the Chuckchi Sea. It is hard to imagine such a remote destination, and it's stranger yet that it was closed to the world by a political door. This is beyond the edge, an adventure in a life of adventures. These words are written quickly in hopes of remembering this scene to communicate to others the experience and reality as we hang on. I told Shane and Gerald that they are great guys and that it has been lots of fun, as if to say good-bye.

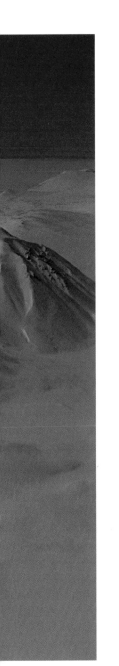

The flight was about three-and-a-half hours from Providenya to Cape Schmidt. As it turned out, a fuel shortage had been one of the many problems adding to our layover. We had to leave by 2:30 p.m. from Providenya because the airport at Cape Schmidt closed at a reported 6:00 p.m.—we were off that schedule by seventy minutes. To this point, we don't think about individual situations or single problems in Russia—they always factored in multiples.

The heater in the plane did not work and the mechanic Vladimir shrugged that off with a head shake and a "nyet." We all had every bit of our cold weather gear on (luckily we were planning for survival in Arctic conditions), but everybody mentioned being numb in the hands and feet. Shane reported later that he thought it was at least minus 20 or 30 degrees. I hoped the pilots' heads were still warm. The long flight allowed the experience of using the bathroom. We first noticed Vladimir talking to the pilots as they adjusted the angle of the nose to compensate for the weight in the back of the plane. When it was my turn I could not find the toilet and had to be shown with Russian language being yelled over the engine noise. A funnel existed in one corner at waist level with a suction sound from the air outside.

Flying past the mountains, we reached the coastal plain. I was motioned to go up front. What a 180-degree view! The pilots put a red bar (utilized as a steering-wheel lock when the aircraft is parked) behind me as they pushed me forward. The bar fell into place in the entry aisle-way between the pilot's seats, then an aluminum notebook was placed on top of the red bar and this arrangement was indicated to be for my use as a backrest. Big smiles came from the pilots. Then I realized they had just a little heat coming through from the engine. It wasn't a lot of warmth, but as I looked back toward the rear of the aircraft I saw that I was the plug that kept the heat forward on the pilots.

For the moment, I was in the best seat in the plane and at least the front of me tingled with a partial thawing. The instruments were all labeled in a Russian script of mixed letters and numbers, but familiar in layout, like a DeHavelyn Otter's Canadian-made instrumentation. Using a slide rule, something I haven't seen since grade school, the pilots nervously argued, pointing at the map and tapping a gauge. When I pointed, the pilot said "Kilos of fuel" in good English. My first words must have sounded like a chipmunk chattering, as I asked many questions at once. He replied, "Not vvverrryyyy much English, low on fuel."

The next situation began from a radio transmission. A three-way yelling, with banging on the steering wheel between the two pilots and a crackling voice miles away. I threw up my hands when Roman said "Airport closed." Much tapping on the gas gauge, now reading less than 100 kilos on a scale of 100 to 1600. We had been flying at about 6,000 feet in altitude.

The pilots pushed me forward to yell back at Vladimir, making my ears flinch. The three other passengers, Shane, Gerald and Misha were directed into a squatting position, leaning forward and braced. The pilots tightened their seat belts and harnesses. I stood up to move back and was pulled back into my teeter-totter stance by Mikhail.

The aircraft began losing elevation and not knowing Russian, I was a bit confused. I turned to Shane and said, "They are low on fuel," then Misha added, "They will try to conserve fuel by flying at a lower altitude with all the weight forward." The airplane leveled off at 700 feet.

It was 6:50 p.m. More radio messages came in, followed by sharp responses. Misha added from the back, "If we have to land someplace else we need to put more gear forward." Shane moved gear with Validimir, then leaned forward on top of the growing pile.

Arguing between the pilots broke out over the slide ruler and map; now the red light of the fuel gauge was solidly on. Tapping the instrument, Roman switched from the right-wing to the left-wing tank. The instrument did not register to the first level mark and occasionally the red light blinked.

We had flown for thirty minutes with red lights flashing low fuel levels. The pilots acted like they were on pins and needles. A new argument broke out between them over where on the map we were. One pointed, then the other shook his head muttering "Nyet, nyet" (the only Russian word I knew, meaning "no"). Then they

pointed on the map to a tip of land with a little airplane next to the ocean. After being in the "crash position" for nearly an hour—an eternal hour—we finally landed.

The pilots in unison blessed themselves and mumbled what appeared to be a prayer. Then Roman looked at me and said, "No fuel; lucky."

After meeting the next day with the territorial governor to get permission to visit as biologists, we went to the airport to load our gear into another AH-2 bi-wing aircraft. This one had skis instead of wheels. After a false start, we discovered that the ground crew forgot to take off the oil-cooler intake plug. We restarted and taxied down the snow-field. A two-hour flight took us over the ice-covered Chuckchi Sea, with occasional narrow cracks of open water. As we closed in on the cliffs of Wrangel Island, I welcomed the topographical relief, after the vastness of the open sea ice. Within the cliffs lies the community of Ushakovska, the only village on Wrangel Island. The rest of the island is a nature reserve, not open to the public. Surrounded by remote Arctic wilderness, the small settlement seemed upbeat and refreshing after the depressing surroundings and conditions of Providenya. We were made welcome as guests, for with the arrival of the plane came supplies and mail.

Over dinner we met with the Park Director Dimitry Pikvnol and our guide Pasha. We made plans to go to Warring Point on the northeast end of the island. We would travel over the sea ice and island by *booran* (a one-skied snowmobile with two rear tracks), and a military half-track that had an old machine gun mount over the mid-engine (the gun had been removed). We ultimately just called this vehicle "the tank."

Dinner was prepared by Pasha's wife. Pasha and his wife were partly of native Chuckchi ancestry. She worked as the village's meat grinder, and she prepared ground caribou patties for dinner. Spaghetti noodles, fresh bread and the usual vodka and cognac were served. Vodka was traditional. A special offering of Kavas was poured from a recycled plastic jug. This fermented black bread, sugar, yeast water and a few raisins produced a soda with a punch!

The next day we prepared for the two and a half hours it would take to travel the fifty kilometers across the island to Warring Camp. Shane and I were stuffed in the

back of "the tank" with our gear, again in one of those uncomfortable situations, while Pasha drove the booran. The noise was deafening. Diesel fumes found their way through the vehicle's many cracks, along with fine powdery snow. Within minutes we were covered in snow, dizzy with carbon monoxide and shuddering with the vibrations of the tank's movements. Over ice and other obstacles we bounced and shook on our makeshift seats atop our gear pile. The only two windows were six inches by twelve inches and were covered with ice and snow so thick that only a tiny bit of light shone through. We could use a red button in the back to signal the front, as a sort of alarm. After an hour of this, Shane was a pale white and he suggested we stop for a minute (and as one who really never complains, this was significant). I pressed the red button and we drew to a halt.

Shane and I crawled out of the back, staggering in waves of dizziness and trying to breathe in the fresh sub-zero air. Later Shane admitted to a severe headache. Reluctantly we crawled back in, and this time I went to the forward position and found a vent window to open. At times snow blew in too hard so the vent had to be closed and monitored, but the air circulation was improved.

The cramped quarters did not allow for free movement, and soon we gave in to cramped chills. Cold, dizzy, vibrated, deafened, cramped and nauseous, we arrived at a small military radio trailer left over from World War II. The trailer measured six feet by ten feet; it had a tiny attached entry room with a snow floor. This cold-room entry had some space for storage and served to buffer the trailer from the outside air when entering. The doors had been broken down and were lying on the ground surrounded by polar bear tracks. The entry and trailer had snow drifts and needed to be shoveled out. The small size of the trailer meant there would only be room for two to sleep inside, so I opted to bunk in the entry room.

In the hard blowing winds, we set up a quilted "yurt-cover" over the structure, and a water-resistant cover over that insulation. The materials were designed for a stand-alone rounded tent, now modified to help enclose and protect the hut. The cold weather was hard on bare skin and our hands froze, to the point of pin-prickling cold immobility. Even with our hard-working movements, heat dissipated at a rapid rate. The wind imposed a deep, penetrating chill.

The tank crew left Pasha, Shane and I to the vastness of ice in the heart of the land of the polar bear. I sighed in weary appreciation as Pasha told us he had spotted a polar bear den across the valley, one-third of a mile from our camp.

Map labels:

WRANGEL ISLAND, RUSSIA

420
396
601
854
1096 m.
420
408
411

Tyîke (Pilar)
Warring Camp
Ooshakouski
Somnitelnia Camp
Blossom Camp

180°
178°
71°

N

Warring Camp Discovery— Ferosha

WRANGEL ISLAND IS SHAPED like the profile of a bear skull. The top of the island (or bear head) is north, the nose faces west and the spinal cord and vertebrae attaches on the east. The first vertebrae in the spine is called the atlas vertebrae, and attaches to a hole in the skull at the occipital bone. It is on the top of the occipital lobe of the bear skull map of Wrangel Island where Warring Camp is located.

It was mid-afternoon and plenty of daylight remained. Instantly Pasha started up the stove and we tackled chores. The three of us first worked on securing the door. Then Pasha dragged in a caribou carcass he had brought for our meals and using an ax, chopped off a leg to cook for dinner. Shane and I readied camera gear and packed the remaining gear into the entry room. After other sundry chores, we all admitted to a desire to investigate the den site across the valley. Pasha drove the booran; Shane and I walked the short distance.

After seeing fresh tracks we approached nervously. With a meter-long flat spade, Pasha broke a chunk of ice from the den entrance and tossed it into the hole. Moving in closer with another piece of ice, he threw it down the tunnel. Much to our amazement, a female bear growled and poked her head out. Getting within a few feet of a polar bear, this definitely was not part of our game plan!

We had not prepared for this. What were we thinking? We didn't have flares, or a gun, and our cameras weren't set up for action. We weren't even mentally prepared to respond—we were defenseless. The bear retreated back into the den, and we heard only a few low-pitched moans.

Pasha moved a hundred yards off and began molding snow and ice into cubes to build a snow blind. Shane suggested a better location at an angle good for filming. Pasha retreated to his booran and retrieved a small hand-saw. He cut wedges of ice and the three of us erected a great dual-ported snow blind. We cleared and simultaneously dug the blind's floor down by the removal of the blocks of snow, as the snow blocks were cut and erected around the perimeter. To build this encircling shelter took about forty minutes. We then left to lessen any further disturbance to the area and let the den occupants relax.

Shane and I took a walk as the first night's sun dipped below the horizon near
midnight. We headed toward the cliffs overlooking the sea. Shane spotted a mother
polar bear and a single cub on the northwest ridge headed toward the sea, about a
mile out from our location. Again we were without flares or other protection,
beyond a couple of cameras. Shane retreated to the cabin to get his camera, and I
just couldn't help following him in lieu of standing out there alone with polar bears
running around.

From closer to the cabin we watched as the polar bear followed another ridge down
to the sea ice. The cub was so small that its detail was lost in the distance. Out
amid the blue pack ice they disappeared, replaced by a halo, formed by the rays of
the setting sun surrounding the ridge where they had recently passed.

This close-to-camp bear viewing lifted our spirits with expectation and high potential—a welcome relief to the ten days of travel. As it began to get dusky, we retreated to the hut to prepare our own dens. I noticed Pasha had hung the caribou carcass just above my sleeping area in the cold room. I added up an equation of concern: a door broken in by a polar bear, a polar bear with a cub running around the neighborhood, a fresh caribou carcass near my sleeping area, the door to the main hut closed to conserve heat. I had an uneasy feeling as I lay down to sleep. After a couple hours of tossing and turning I decided maybe something else was causing my sleeplessness. I didn't have to dress to step outside to have a kidney break; I was wearing everything I had brought, plus my sleeping bag. When I pushed my way through the yurt flaps immediately in the night sky I saw the "dance of the bears in the sky."

The northern lights moved in the north sky against the constellation Ursa Major and Ursa Minor. Against the snowy landscape the lights seemed surreal. I lost my orientation as the lights moved quickly above the landscape. In the chill, I watched for almost an hour, as I finally felt myself truly arrive at this remote place in the middle of the night. My body caught up to my mind, repairing the separation that occurs from so much travel. I'd been grounded by the indescribable phenomenon of the Aurora Borealis. This awesome spectacle rendered a very spiritual tone in me, touching my higher spirit with a supernatural feeling.

<hr />

The next day came early, and with sharp anticipation we made plans: Shane would set up cameras in the snow blind at den number one. Pasha and I would do reconnaissance and hopefully find more photographic options at other dens. We hooked up the sled that had carried the caribou carcass, and I loaded my camera packs and survival gear onto the sled. Pasha drove the booran.

Pasha and I traded driving and covered seventy kilometers within which we located two more den sites. One den had fresh tracks where the mother had led two cubs to the sea ice. A second den, number two, was within a few kilometers of camp. We decided to approach it later so as not to disturb it; there were no tracks nearby implying it might be occupied. We purposely kept our distance and monitored it from afar.

Shane had returned to the hut. He had seen fresh tracks from a mother and solo cub that had been left along a ridge. Shane had followed these tracks and decided they had been from the bears we had viewed the night before. Disappointed that the bears closest to camp apparently had left the area, we discussed strategies.

One of the many occupational hazards of being a nature photographer is being hit with the phrase, "You should have been here last year, week, month or yesterday." Pasha wrote down the following:

1991, 27 Bear Dens

1992, No Research

1993, 41 Bear Dens

1994, 8 Bear Dens

Apparently not only was this a non-peak year, but the ten-day delay in getting here could have put us in too late—when most of the females with cubs had left their dens and were now out on the sea ice. Pasha's dictionary English claim is "Bear tracks all lead to the sea ice, last year better, earlier best."

After a dinner of fried caribou (last night's dinner revisited), we explored the den that we called number two. Shane and I set cameras on tripods ready to shoot, and carried them atop our packs. Pasha led the way as Shane and I followed toward the unexplored den. Female polar bears build dens in wind slopes where snow accumulates to depths suitable for a tunnel and an insulating den.

We were walking along a crest of a half-moon-shaped ridge five to six hundred feet in height. The ice snow was so hard our footsteps didn't leave much of an impression. Carefully balancing our backpacks, tripods, cameras and long lenses we moved in single-file along the ridge.

Suddenly, a couple yards away, a polar bear's head popped out of the snow.

Slightly stunned, I took a penetrating look around—permanently burned in my mind is the uphill lip that surrounded the den entrance. From then on I knew where this den entrance was. Pasha tucked his legs up and slid down the slope and instantly was out of sight. Unable to follow with our cameras and packs, we

carefully turned on the icy ridge and breathing heavily, backtracked out of the den site. Taking a long route around we met up with Pasha, and in Russian agreed about *"Bearloga medvid havasho"* (den polar bear good).

Shane selected a good snow-blind location, and this time we constructed an even more elaborate blind for den number two as the snow was deeper and more solid than at the first den. Hope returned that photographic results were possible as we headed back to camp in the twilight hours. Rose-colored snowy hills outlined a red cloud in the northwestern sky. Back at camp we settled in for sleep with the next day's assignment prepared. We relaxed after the excitement of meeting polar bears in unexpected close proximity.

⊱─━◆─○─◆━─⊰

The next day was another lucky weather day, clear and cold. We dropped Shane off near den two. Pasha and I scouted for more dens taking turns driving and checking for tracks. In the course of the morning scout we only found a few tracks. Pasha showed me around through a walk on the sea ice among the icebergs, pack ice and the cliffs above. The 2,000-foot-tall shale cliffs are used in the summer by nesting bird colonies.

About two in the afternoon while scouting in the booran, Pasha spotted a polar bear. He stopped and pointed as I prepared cameras and a 500mm lens on a tripod. Carrying the camera gear in ready position on my shoulder, we approached from around a concealing ridge, downwind. Pasha stopped again then began walking and signaling to me to follow. He was able to proceed more quickly than I could with my gear. He scouted over a ridge by peeking his head over a snowdrift, then he signaled "down and to the right." I caught up to him by taking an immediate right turn, then walked a few hundred feet to see a polar bear and her cubs playing in the snow 75 to 100 yards away.

I squeezed off a few frames while I watched, documenting their play in sliding, rolling and various mother-and-cub interactions. The wind must have shifted, for within minutes the mother snorted and ran to a double-entrance den. One entrance was below the snow ridge, the other on top and six feet back from the edge. The sow went in first, almost in a dive, and the cub looked around a couple times and quickly followed.

Where we stood wasn't a good site for a snow blind, so we retreated to the booran and detached the sled. We'd left the shovel and saw at camp, so Pasha sped off on the booran as I scouted a good location to build the blind. The den was located at a headland where a creek canyon begins to widen to the coast. In all the excitement we must have lost our minds. I lost sight of Pasha, last seen in my binoculars going over a ridge. I was without any defense and without additional cold-weather gear. I pulled the sled by hand along with me until I was level with the den, across the frozen creek.

This natural snow-blind site was in direct wind. It soon shifted direction, and began to blow snow. The snow mixed with wind swirled to a more rapid pace and blew down the canyon. After half an hour I began to chill and when an hour was up, I was in the early stages of hypothermia.

With my tripod I broke chunks of snow and dug in to try to escape the chilling wind. I tipped the sled over to further block the wind. An hour and a half passed and no sign of Pasha. I now had a bad headache, endless shivers and the numbness in my hands and feet was becoming very painful.

It was two hours before I heard the booran again off in the distance. When Pasha finally arrived he asked about the *bearly midvid* (white bear). I had forgotten about the bear, and remembering added to my already considerable anxieties the sense of having been alone with polar bears one hundred yards away. Pasha turned to me and in English pronounced "long trail to camp." I sat up on the booran close to the motor and warmed myself.

Automatically we connected the sled to the booran and loaded up, then Pasha cut some block ice and marked the snow-blind site. Soon we were off as we tucked behind a different ridge this time out of the prevailing wind.

When camp was in view across a long valley and a ridge, Pasha stopped and was checking my face for frostbite, until he saw my eyes bulge at what I saw behind him.

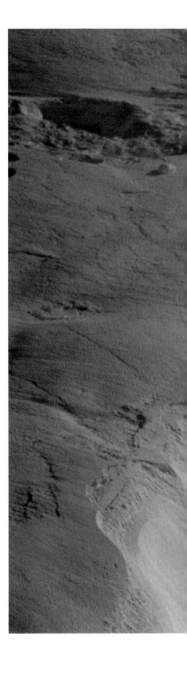

I wasn't sure Pasha saw it and I couldn't really communicate well enough except with "bearly midvid."

Pasha did not have goggles or sunglasses, and although he was an incredible spotter, he missed this movement nearby. A tiny polar bear cub was less than fifty feet away, next to a block of snow watching us. My first reaction was to wonder where its mother was.

When Pasha did finally see the cub, he became jumpy and began to quickly look about. Then he ran up to a ridge that sloped to the sea ice that blocked the north horizon. I broke out my cameras and began shooting this unbelievable opportunity. Soon the 500mm lens was surrendered to an 80–200mm lens, and when the cub approached closer, I attached a standard lens. Pasha patrolled out there walking alone, looking for a female polar bear, until he felt the region was clear. I had been left with the booran, which I could start up if the female arrived or popped out of an undisclosed den.

When Pasha returned he stated, "*Nyet Bearloga midvid.*"(No white bear den.) Then he pointed out a word in the pocket dictionary: "orphaned." With the cub now close by, we could tell it was either suffering from the elements, or scared, or hungry, or all of the above. Both of us were also cold and with camp within view, we decided to retreat.

As we readied the sled to head home, the cub rotated its tiny head watching us as we moved about. The two- to four-month-old cub stood about ten inches high and eighteen inches long. As we pulled away the cub followed. Pasha stopped and picked it up by the excess fur on its back and placed it in my lap on the sled. Saying nothing, we were soon back at camp and inside the hut with a polar bear cub!

By the time I began thawing (after aspirin, warm tea and a vitamin), I felt intensely ill. Pasha asked if I had aspirin for him, too, as we drank tea together. I lay down and within moments fell asleep. The cub and Pasha took an hour nap as we warmed by the stove.

In retrospect, I was dehydrated, hungry and tired, and because of the wind chill possibly the coldest I had ever been. The dehydration, dense meat (a change from my usual vegetarian diet), and cold had not allowed for my normal body functions for a few days. The aspirin did not relieve the headache so I forced myself to drink tea, cup after cup, to rehydrate.

After an hour we were due for our rendezvous to pick up Shane. The wind had increased to a white-out. Shane had begun walking back earlier than the 5:00 p.m. scheduled pick up, and we met him at the top of the south ridge. We did not say anything, as the wind limited communication. Back at camp, Pasha and I, aware of the cub, watched as Shane entered the hut first.

The day had been rough on us all, and Shane didn't immediately realize that there was a polar bear cub in the hut. Pasha had already cut up caribou and started cooking it when the cub woke up from a nap near my sleeping area in the entry room. Shane didn't say much about this unprecedented event. Neither the uniqueness of the cub's presence nor the same old dinner menu could arouse any further reaction. Pasha picked up the cub and pronounced it female. He named her Ferosha, which in Russian means *ferocious*. We all ate and quickly drifted to sleep.

Mixed emotions mingled in my mind as I pulled the sleeping bag around me. Nature should run her course—the cub should not have been introduced into our world. A situation like this has never presented itself to me before. Instead of a decisive direction, I went along with the flow. I watched in dim light as the cub curled up only a few feet away and imagined what her life had been like before our encounter. Where were her mother and siblings, if any? Which den did she come from? What was to become of her?

A decision to let nature run its course would, however, be a certain death warrant for the cub. Without her mother's parental care over the next two to three years she could not survive. During the springtime, polar bears will mate if circumstances are right. During the summer the fertilized egg goes into a delayed implantation, waiting until the late summer to implant onto the uterine wall. In late November or December pregnant females find snowdrifts onshore to dig a den in a traditional denning area such as Wrangel Island. Some dens are made on pack ice, usually in reaction to onshore disturbances by people and industrialization, as is the case in Northern Alaska.

On average, two cubs are born in late December or January. One to four cubs is the range, with four being extremely rare and three cubs uncommon. Cubs grow rapidly: at birth they weigh only about a pound, are blind and have very little fur. In three months the cubs grow to about fifteen pounds and are ready to leave the den. At this point they are tough enough to face a day like the three of us experienced when we found her, and possibly worse, as the sea ice holds many more hazards.

I wondered about the cub's mother. Was this cub a runt? Was the mother young and unable to nurture this cub? Or was this cub just not able to keep up with its mother on the way to the ice? I finally fell asleep closer to polar bears than I could have imagined, physically and mentally.

I must not have been asleep long when I was awakened by the cub bawling. When I sat up the cub placed her paws on my knee, looking me straight in the face. Her button eyes pushed the ancestral nurture button. I had heard this call before—

it wanted mother's milk. After the fright of waking from a sound sleep, I lay back down and the cub pushed between my knees and curled up to sleep. Every so often through the thickness of my sleeping bag and all the clothes, I could feel the gnawing and pawing, looking for milk.

With this new mouth to feed, I was glad that Pasha had removed the caribou carcass from the entry room, my bedroom, and now polar bear cub den. He thought it a good idea to place it right outside the door, hanging a few feet away. But in retrospect, perhaps this wasn't so wise. Camping in polar bear country, bears knock down doors even when there isn't a caribou carcass beckoning to enter. Here we had a carcass flapping alongside the door entrance and a bawling polar bear cub inside. No guns, just a couple of flares and a snow shovel.

Fortunately, during our time at Warring Camp no bears came calling in the night.

A Dream Realized

LONG AGO I HAD LEARNED that wild bears are individuals with unpredictable behavior, but on Wrangel Island I observed their curiosity was more focused on hunting seals out on the ice than on investigating humans on the land.

It was my dream to be present when a mother polar bear led her cubs out of the den for the first time—to capture on film the cubs' behavior as they got their first glimpse of the world waiting for them beyond the den.

Seal-pupping time was upon us, and part of the reason mother polar bears bring cubs out of the den at this time is the abundant food on the ice. The ringed and bearded seal pups are the single most important food to polar bears. When seal pups are about a month old, they're fat enough to present a tasty meal for polar bear.

The importance of the timing of this bear behavior did not make sense for me at first. But then I realized these bears focus their feeding efforts on seal pups and patrol the ocean's pack ice intensely at pupping time. Springtime prompts peak feeding activity in polar bears and is a critical time for their overall survival. It was nearly impossible for me to get close to a hunting bear for a photo. One difficulty was that they patrolled an extremely large area. Secondly, the floe edge, with its tall fissures and rifts of pack ice (along with cracks in the ice itself) limited my access to the range the bears covered.

Tuesday, April 12

Before 7:00 am we were up and on our way to den number two. We left the tiny orphaned cub in the warmth of the Wrangel Island hut, sleeping soundly as if hibernating.

Fog filled the valley as the sun began to rise, developing a fog bow in a circle around the sunrise. We built a small blind, less obtrusive to the nearby den. We placed it about 150 yards away from the den entrance, a bit too far away for still photographs. I waited the morning there with Shane, but we didn't see any bears. In the afternoon Pasha returned on foot with news about another new den location. We hadn't had any visual bear contact at this den site, so I decided to go with Pasha to set up a snow blind at den four.

I carefully gathered my gear for the walk back to the junction of the snow trail, about a half-mile over a ridge. The new den was another mile from the junction, toward the east and closer to the sea ice. With my eighty pounds of camera gear I heated up on the trip, growing clammy inside my layers of clothes. As we arrived near the new den, I picked a spot just to the edge of a south ridge, with the den on a south slope one hundred yards away.

The approach to the snow blind site was out of view of the den entrance, and with a prevailing easterly wind our scent would stay away from the den as well. Pasha and I built the most elaborate snow blind to date. It had room for two people plus

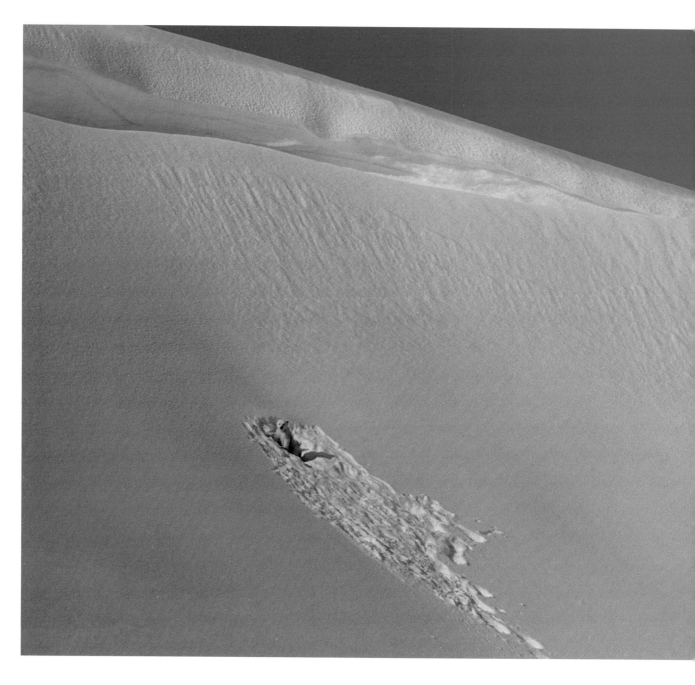

gear, and had wraparound protection from the wind. I had bought bed sheet
material when I was in Nome, and used some cut pieces to cover the camera-lens
ports. This total concealment needed a couple peek holes on the sides to watch in
the other directions for any other approaching bears. We sawed some thin,
elongated view-slots in the ice walls. Pasha left and I completed the blind work.

This spot was a great site: the bay was close by, and we could hear the thundering
and cracking of the moving sea ice. Later in the afternoon, the shifting tide
beneath the ice pack created a wide crack, sending a cloud of fog up from the newly
exposed open water.

At 5:00 p.m. the sun passed beyond the ridge and the snow blind and the den fell into a deep shadow. The offshore breeze quickly turned into a storm wind that was building in strength. After an hour, when blowing snow obliterated the den entrance, I retreated toward camp. Walking back against the wind is a classic Arctic experience. This storm was brewing up to be a big one. Halfway to camp I heard Pasha on the booran drive by, but I couldn't see him. I had been following a different snow-machine track. He was en route to pick up Shane at den number two, and I reached the hut just as they arrived.

We couldn't attempt any photographic work for the next two days; visibility was near zero and the horizontally blowing snow made even everyday chores a challenge.

▸━◦━◂

Thursday, April 14

Day spent in the hut avoiding a wicked storm. The wind is blowing out of the northwest at 20 to 30 mph with gusts to 40 mph. I can feel every incremental increase in the speed of the wind as an intensifying cold. Not much to convey about cabin life, although inside a 10 x 6 foot hut one can only imagine how crowded three guys are all day. Compound this with the language barrier and we have an interesting combination. We all spent time playing with the polar bear cub.

Late Friday afternoon the storm subsided and the four of us gladly left the hut. Pasha, Shane, Ferosha and I hiked over to den number one. I resolved to take some photos to make up for the lost days from the storm and those missing days due to the difficult travel. Ferosha was quite energetic, tough and feisty. When we arrived at the den, no new bear tracks were visible; much was concealed by the recent winds and snows.

The den entrance was narrow; it was hard to imagine a polar bear moving through it. With my hands stretched forward I pushed my camera ahead and inched slowly inside, pushing with my toes and lifting with my forearms. The entrance tunnel was a dark gray zone, twenty to thirty feet long, with a curve toward the right so I couldn't see the end until nearly upon it.

As I rounded the curve I saw that a sunken oval shape about the size of a bear was at the far end. The den itself was an open, glacial-blue cavern, brighter than the tunnel. On the roof were patterns of claw marks revealing bluish-colored snow, as if the bear resting on its back had scratched the condensation that gathered. The scraping had thinned the den's ceiling to a very sparse layer of snow, which allowed the bluish illumination. A layer of ice coated the nearly perfectly smooth surface.

Even now my breath made a newly forming visible moisture layer.

Crawling in, I passed over excrement and a few urine spots. Inside the den there were a few patches of white hair and fuzzy fur stuck to the interior of the walls and on the snow. Ferosha followed me inside and seemed to be at home. She sniffed at the excrement and even tasted a small bit, and rubbed against the main chamber wall. I was lying on my back to photograph the ceiling with its distinctive claw marks when the cub stepped up on my chest and lay comfortably down.

For these moments my world spun. It was one dream to locate a bear den, yet another to crawl inside of one and witness a part of a polar bear's life that is seldom seen. The cub's presence brought together these awesome events with a touch of reality. A den like this was the cub's true home; her life cycle passed through my mind. Her mother, the birth, her development safe inside a den, and eventually her departure from this safety and toward her unknown future. The first look at the huge icy realm outside of her den must have been a grand revelation.

It was quiet inside the chamber. I couldn't hear Shane, Pasha or the wind. Shane crawled part way inside to say they were freezing and were ready to go. I took a few photos with a snow-covered camera and exhaled to squeeze back out.

In the middle of my tunneling, I heard yelling; from above the snow the sound was muffled and as if it had come from a long distance. I had a claustrophobic moment, fearing the worst. Was the mother polar bear returning? Was something wrong with Pasha and Shane? Anxiety built. The final five feet of the tunnel were tricky, a dog leg up-and-forward bend. I could hear myself pop as I exited, like a cork out of a bottle.

A resurgence of the storm with great wind gusts had developed, and Shane and Pasha had to yell to be heard. We headed to camp. Ferosha followed after a little play in and around the den site. The sun was setting as we arrived at the camp a few minutes later. The night's red-orange sky color was spectacular against the icy landscape.

That evening something inside of Ferosha must have turned her behavior. The tiny cub moaned and cried all night. In her ramblings she left a dropping on Shane's duffel bag, spilled over some kerosene onto my sleeping bag and was a manic terror well into the wee hours of the morning.

Saturday, April 16

After two days of stormy weather with a clearing on the way, we needed an early start. We headed out to blind number three. Waiting in the blind the next morning, I twice lost movement in my toes. Both times I carefully exited the blind and crawled over the ridge, where I did a hundred sit ups, jumping jacks and other stretches, running in place and exercising until circulation was reestablished and some heat returned to my extremities.

Pasha claimed that it was minus thirty; with the north wind at 5 to 10 mph this dropped the wind chill to minus fifty or so. Pasha had come to tell me that a helicopter had landed to take us out on the sea ice with researcher Gerald Garner. Garner had been stuck with all of us in Providenya, and was stuck for two more weeks in Ushakovskoe village waiting for weather and a helicopter.

Today was his big opportunity to place radio telemetry collars onto polar bear mothers with cubs, the focus of his research. Along with him was Dimitry, the Wrangel Island Reserve Director, biologists Michail Stishov (Misha) and his wife Irina E. Menyushina (Travina), and their eight-year-old son Ivan.

The team was to tranquilize a mother bear from the air, then land and place the radio collar, tattoo the bear's lip for permanent identification, take blood samples, run an electrical current test to measure the fat reserves, pull a remnant tooth for ageing, take a hair sample for analysis, do an overall condition check and a parasite review. They would also weigh, measure and review the health of the cubs.

Gerald exclaimed that he was glad to be out working as he stepped out of the helicopter. "The bad news is that we have found only one bear family all morning. Typically we see bears all over, just offshore," he said. He asked if I was ready to go, but I needed to get different camera equipment and more film, so I ran to the hut.

Pasha had Ferosha in his arms and was heading toward the helicopter with Dimitry. The decision had been made by Dimitry to take Ferosha back to Ushakovskoe where proper care and a future home could be organized. I hurriedly gathered my gear for the sea-ice aerial.

As I returned to the large orange Aeroflot helicopter, I was amazed by all the space inside. Pasha was saying good-by as he stepped off. Ferosha lay between the legs of Irina and Ivan. We immediately took off over the frozen Chuckchi Sea.

I stayed pressed to the round window of the helicopter as I surveyed the land I had familiarized myself with over the past weeks. There was the pillar which had become a familiar prominent feature, the cliffs, den number three and further out lay new territory of crack ice, floe edge and pack ice. I felt like a kid on Christmas Eve. In my excitement I took photographs of the inside of the helicopter, the pilots, the cub in Ivan's lap, and made plans to cover the research in progress. The ice pack was interesting too, and I composed some aerial photographs.

Everyone played with Ferosha as she roamed free and seemingly at ease inside this strange new environment. Nothing was said about her or what plans would be made for her, and I was hesitant to bring up the subject. Ferosha stepped up to the cockpit and had a look around. Gerald was so engrossed in filling out data sheets and preparing for the next round of darting calculations that he didn't take notice of this bear cub now lying with her head upon his toes.

Rooting about in my pockets for more film I came across a candy bar. I passed it over to the boy, Ivan, and his eyes opened up to his hairline. He didn't reach for the candy; instead he looked over at his mother three times. She spoke softly to him, and he smiled and accepted the bar. He passed it to his mother—she looked it over in detail and put it into her pocket. In Providenya we had seen candy bars with a remarkably high price, even with our exchange rate. This was another grounding reality as to where I was and how different things are throughout the world, and in this group of people. A group brought together by an interest in polar bears.

We flew for an hour before we spotted a polar bear sow and two cubs. Near an open
lead, Gerald instructed the team to wait until she moved past the open water, to
avoid having her enter the water after sedation. This allowed for some extra time to
estimate her weight and prepare the immobilizing rifle. With the side door open,
Gerald watched the mother and cubs until a good shot was available. *Pop*.

From a .22 caliber blank the dart flew quickly and connected with the rear flank of the mother bear. A good shot. While Gerald watched, the bear continued traveling as if nothing had happened. After a moment he prepared another dart with a light load of immobilizing drug, which was kept ready in case her weight had been underestimated.

Her cubs kept close beside her—it was amazing to see how much ground they traveled in so short a time. The bears' movements were efficient: front paws outstretched and rear paws kicking out in a sprint motion. A snowmobile would be hard-pressed to match their speed and agility. In a heartbeat the polar bear family sprung over some piled pack ice that would stop a snowmobile in its tracks.

The female bear wandered toward the island from the sea ice, when the drug began to take hold. Her hind legs first began to show the drug's effects. Slowly she pulled with her front paws and dragged her rear along. The helicopter landed nearby. Now the team, organized with gear and assignments, headed in—Gerald with dart gun, Dimitry with shotgun and I with my camera. I was wishing that these guys were beside me in all my work when I approached bears.

The sow had gone into a sound sleep, lying down prone, the cubs touching her body with theirs. As her muscles relaxed she slid backward a bit upon the slight slope where she was resting. This movement brought the guns to eye level, but only the camera fired.

When we approached closely the cubs moved off. Dimitry placed the shotgun upright in the snow and chased after them. The helicopter was back in the air to spot any other bears and keep an eye out for another bear family to track. The team intended to place seven collars, and this was just the second bear this season. The cubs were precious to the project, so the helicopter wrangeld the cubs to a stall until Dimitry caught up to them. They had scrambled almost a mile before they were retrieved. Placed back aside their mother they seemed to become more curious about all the details of the science. When the mother's blood sample was taken, one cub became particularly problematic. A light sedation was given to the cubs to enable samples to be taken from them as well.

Seeing the cubs lying alongside their mother asleep, I could not help but envision three cubs instead of two. In a last hope to free Ferosha, I queried the reserve biologist Misha. Misha was our interpreter, with distinctive, heavily accented English. I could tell he entertained the idea for a moment, he did not respond immediately. Gerald had wires attached to the female bear measuring her fat content and a G.P.S. device to receive our latitude and longitude position. Misha carefully chose his words. He said, "How should I say it, she will kill Ferosha."

Well, that popped my bubble of an idea. So far all the options for this cub pointed to death. Left alone, death from exposure. Left in her den, spared from the cold, death from starvation. The long shot of joining another wild family really meant the female would kill her. It was now obvious that Ferosha was smaller than these healthy-looking cubs. (Unbeknown to us at the time, she would eventually find a home at the International Bear Foundation's Ouwehans Dierenpark Zoo in Holland, and her survival would be assured.)

The team completed their work quickly and efficiently, all working together. Even Ivan, eight years old and the youngest team member, played a helpful role. I kept thinking that in the United States, liability worries would have kept this youngster's experience from ever happening. What a great story for young children everywhere, about the life of a young boy learning the role of a polar bear biologist in Russia. This child was tough—the wind and cold took a toll on all of us.

My cameras played out the Murphy's Law syndrome. At this site, as experienced at intervals throughout the cold-weather trip, the following complications popped up: frosted fog on the back of the camera due to my rapid breathing after running with or after bears; film broken inside the camera under the pressure from advancing or manual rewinding; routine battery chilling and failure; lens frosting over; and difficulties in loading film in minus thirty degrees compounded by blowing snow, wind and the adrenaline of being close to large carnivores.

⊱━◈━○━◈━⊰

Back aboard the helicopter, the report was that no other bears had been found. We flew the remaining afternoon covering the entire northeasterly edge of Wrangel Island over the land and sea ice. Gerald was pessimistic and out of character. When we landed in the evening back at Warring Camp the mood had not changed. Not even Ferosha's antics could lighten it. Pasha made tea and coffee for the crew for a break in the hut. The two pilots and engineer, Dimitry, Misha, Irina and Ivan, Gerald and Pasha all sat, practically on top of each other, squeezed inside the small hut. They deliberately laid out maps for a patrol strategy on the route back to Uschakovskoe village.

Before leaving, Gerald summarized the potential for bears, dens and observations as bleak, compared to prior seasons. In his experience this had been an off-year for the total number of dens. He declared that we should have been working here the week we wound up being stuck in Providenya. He thought that we would be lucky to find any more bears still in the den.

To divert any pessimistic thoughts with a distraction, the young boy and I went over some of my belongings. I had a parka that fit him well, but he was incredulous about the gift and took it off to return it. This time I gave the parka to Misha, his father, who accepted the gift and gave it to his son. With profound appreciation, he put the coat on. When his hands went naturally inside the pockets, he found two more candy bars, and he smiled and raised his eyebrows. He didn't share the information about the candy with his family; rather he happily ran outside.

Later Shane reported on the last remaining known den site, number four. No activity, just blowing snow and wind. Concerned, we made plans that Pasha would continue patrolling farther off seeking any additional den sites, while Shane and I would take up vigilant watch on den site four. The short time that the sun now reached the den slope allowed for an opportunity window between 6:30 a.m. to about 5:00 p.m. We tested both edges of the time frame. After that time the shadows were so dark that it was impossible for photographing.

Tracks would tell us if the female leaves the den during the night, a major concern. In addition, Arctic weather is always a limiting factor for us and our filming. Although we weren't certain that the coming weather presented a problem to us by limiting the bears' movements, it has been hypothesized that bears stay in during storms and may create temporary dens when out on the ice. We could only hope that den number four was indeed occupied, and if so, that the bear and cubs might appear while we could film.

With a week remaining we kept hoping to realize the goal of our trip: to witness a mother and cubs exiting the den. The days of this week fell into place as our den-sitting hours mounted. Shane and I reflected on how impatient park visitors are when they schedule a day in Denali National Park in Alaska and see only a moose, caribou, dall sheep and ptarmigan—but do not see a bear or wolf during their five-hour tour. We had devoted a month to obtain one set of photographs of a behavior that we may never witness. Talk about gamblers!

A well-known polar bear photographer, Dan Guravich, had tried this denning photographic experiment a few years earlier and much like us, he experienced many difficulties. At a slide show of his that I attended, he presented a photographic slide of the one image of a bear looking out of a den entry that he was able to take. Summarizing his expedition with these limited results took a stable self image and great confidence. Dan's many years of professional work enabled him to display humor in his presentation.

Here on Wrangel Island I was faced with the likelihood of failure. Shane was also discussing this reality as we chewed on a tough piece of cooked caribou. Then Shane said with some optimism, "But if you do get the photo (and we still have a week) you could have the only one like it." I reached for a freeze-dried cheesecake dessert to lift my spirits on that hope.

⊱┈◈┈⊰

Sunday, April 17

Clear, cold, minus 20 or 30 depending on the time of daylight, no wind today. The female polar bear popped her head out about 8:00 a.m. and again at 8:08 a.m. It looked similar to the only photograph I had previously seen of this behavior.

After we packed up for the day we heard Pasha on a ridge by the sea ice. Below was a sow and two cubs headed to the ice. Shane and I rushed to reset the cameras. Interesting to see bears as first steps onto the ice are made. The cubs look so small compared to the vastness of sea, land and ice.

Monday, April 18

Another beautiful day, clear, cold, out of bed at 5:00 a.m., out the door before 6:00 a.m., set up in the blind by 6:30 a.m. It's cold, minus twenty five or thirty. I lost feeling in both feet, large warmers assisting to no avail. No photographs.

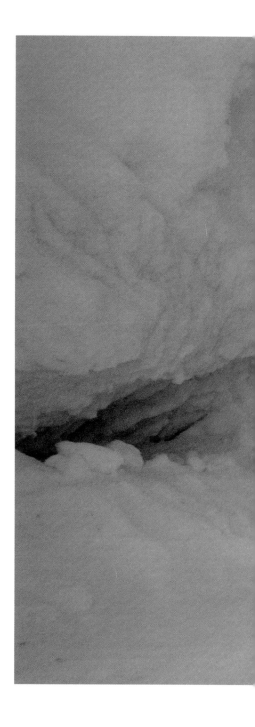

The bear stuck its head out for a peek. We feel the pressure: the last den, two great clear days in a row, who knows? We feel the odds, the bears could come out any time of day or night and leave. We will only know by the tracks. Photographically we can work now between 6:30 a.m. and 4:00 p.m. Between 11:00 a.m. and 1:00 p.m. the light is too harsh for photography. We need some wild luck to achieve our dreams. A prayer is said to nature in hope and faith.

Tuesday, April 19

Lucky, it's clear (and cold) again. Early to the blind. The polar bear female looked out and stretched today. Five or six photographs in the pretty early morning light. The rest of the day is much the same as blind work goes, sit, wait, freeze and finish reading another book.

A snow bunting flew into the blind today, which startled the bird and us. I take this as an indication that spring is coming. The first snow bunting.

Wednesday, April 20

Unbelievable, four clear days in a row.

The mother polar bear peered out at 6:40 a.m., then again at 7:50 a.m., and at 8:45 a.m. a cub came out for just a few moments, and then retreated. Fifteen photos taken, each carefully squeezed off and sound-insulated.

Thursday, April 21

Clear and cold. Early to the blind, no bears seen today. Crystals of ice fell from a clear blue sky. These ice flakes are amazingly jagged-edged and beautiful. Against something dark they measure from one-eighth to one-quarter inch in size and each is geometrically

intricate and unique. We make plans for our departure—a new low point is reached in our disappointment, after the brief glimmer of success yesterday.

Friday, April 22

Clear and cold, up early, no bear activity today. I have become accustomed to feeling cold in my feet; my nose and mustache have had ice on them for two weeks now—even in my sleep. Today I've caught up on my writing, after finishing reading my third book. Shane has read four or five. With only one day remaining I am preparing for the disappointment of not getting the pictures I had anticipated.

I am shooting without a motor drive, as the sound is too loud in the ever-so-quiet snow cover. One frame at a time, saved for just the right moment. The den lies eighty yards away. With weeks of surveying and plenty of time, I have concentrated harder on this one exposure at this location than on any other photo I've previously made.

As I placed the period in the above sentence, my dream came true!

In good afternoon light, at about 3:15 p.m., two cubs appeared at the den entrance and began to play. The largest cub playfully tumbled twenty feet down the steep hard-pack snow, then a third smaller cub appeared. This tiny one reminded me of Ferosha.

Slowly, quietly and without moving, I exposed one frame, one photograph at a time. The cubs played at the den entrance until it was marked up with claws, rolls and digs.

After a half-hour, the mother peered out and the cubs greeted her. She sniffed the air and squinted in the open surroundings. I photographed the scene until film required changing and then I quickly grabbed another fully loaded camera.

Upside-down, a cub slid downhill as its sibling bit at its feet. The small cub decided to explore above the den, taking the first few steps over its mother's head. Our anticipation was to accomplish a photo of the first moments a cub comes out of the den—and here it was unfolding in front of me. It was beyond any expectations.

My feelings went on a roller coaster, from weeks of anticipation, preparation, travel and dreams to witness this young, fresh new life in its unending joyful play, as if in celebration of life.

I will long remember these days, not only for the struggle, but for the lessons nature provided about life. From the bleak, desolate ice cap, this unique contrast.

As we watched, the weather became increasingly stormy. Soon, the cubs and their mother once again slipped back into the den. We, too, finally surrendered to the elements and retreated to our camp.

This was our last day. We would leave knowing that the new family would soon find its destiny out on the sea ice.

Saturday, April 23

Before bed, I went out into the storm and sat in meditation with nature. The polar bear is the strongest example of life I had grown to know. In learning about this magnificent creature I had learned about myself. I envied its abilities and adaptations. Weaker, I am humbled in its presence.

A feeling of overall elation surfaced—I had the good fortune to have learned and witnessed a rare sighting. I will carry the tools I have learned from the cold, but mostly from the polar bear.

Somehow I feel I'm changing: these are the last days of my third decade of life, and I feel gratitude to have seen the polar bear's strength of spirit, in joyful play and tenacity of life.

Epilogue

I TRAVELED HOME TO KAREN TO TELL HER STORIES, share my discoveries and to pass on tales of the wild strength of a polar bear cub challenging the Arctic sea ice.

Spending time with polar bears was first a lesson in survival and living in the Arctic. The habitat, landscape and sea ice is incomparable, and the polar bear is a product of this unworldly, unique niche. The wanderings of polar bears around the Arctic ice cap is unparalleled among the carnivores in distance traveled. The folklore, legends and stories surrounding polar bears are filled with myth, fear and awe.

For me it was the spirit of the journey that persevered through the hazards, failures and limited viewing. But in the end the challenge was an education. I would not have missed any part of it for I have adopted some of the survival skills, patience and understanding given to me from the Arctic and the Ice Bear.

As this trip wound to a close, I found that I am never ready to leave. It helps me to imagine observing bears, and their wonderful world, at another time and place as I plan the next expedition. This always helps—I never really leave, I just borrow moments from the future.